T0353397

Physical Oceanography and Climate

Climate research over recent decades has shown that the interaction between the ocean and atmosphere drives the global climate system. This engaging and accessible textbook focuses on climate dynamics from the perspective of the upper ocean, and specifically on the interaction between the atmosphere and ocean. It describes the fundamental physics and dynamics governing the behavior of the ocean, and how it interacts with the atmosphere, giving rise to natural climate variability and influencing climate change. Including end-of-chapter questions and turn-key access to online, research-quality data sets, it allows readers the chance to apply their knowledge and work with real data. Comprehensive information is also provided on the data sets used to produce the numerous illustrations, allowing students to dive deeper into the data themselves. Providing an accessible treatment of physical oceanography, it is perfect for intermediate to advanced students wishing to gain an interdisciplinary introduction to climate science and oceanography.

Kris Karnauskas is Associate Professor in the Department of Atmospheric and Oceanic Sciences and Fellow of the Cooperative Institute for Research in Environmental Sciences at the University of Colorado Boulder, with a secondary faculty appointment in the Colorado School of Public Health. Prior to these positions he was a member of the scientific staff at the Woods Hole Oceanographic Institution. Kris currently serves as Editor of the *Journal of Geophysical Research – Oceans* and recently served on the Scientific Steering Committee of the US Climate Variability and Predictability Program. Kris was the recipient of the 2017 Ocean Sciences Early Career Award from the American Geophysical Union. He is frequently a contractor to the United Nations Development Programme, providing expertise on regional climate change impacts for small island nations.

'From Ekman to ENSO, Prof Karnauskas covers all the essential classic and modern topics of oceanic and atmospheric dynamics, including their interactive physics from the outset, using a lively style of writing enhanced with compelling graphics that will appeal to a wide range of advanced undergraduates in physical oceanography and climate sciences, as well as to cross-disciplinary Earth system scientists in Ph.D. programs. His writing style is refreshingly accessible through his colorfully expressive descriptions, which are firmly rooted in the indispensable mathematical foundations, that render clear expositions of the key topics that link together the intricate dynamics and thermodynamics of climate variability and climate change. He presents a novel and uniquely integrated perspective of climate variability, ocean-atmosphere interactions, and global warming, vividly illustrated with his self-designed schematic diagrams based on his own vast experience in studying both components of the system and their communication through the air-sea interface.'
– Arthur J. Miller, Scripps Institution of Oceanography, University of California, San Diego

'By standing at the ocean's surface and looking at the ocean through a climate lens, this advanced undergraduate text provides a focused view of the essential place of the ocean in the coupled climate system. With vivid prose and clear explications of mathematical necessities Karnauskas has created an exceptionally efficient means to understand the climate system. Its "deep dives into data" is an appealing feature that makes it easy for the student to explore the climate on her own. At this time of urgent interest in climate this book should find the wide audience it deserves.'
– Mark A. Cane, Lamont-Doherty Earth Observatory, Columbia University

'This text book gives an accessible and comprehensive overview of the processes in the ocean that are important for climate, for upper level undergraduates to graduate students in the ocean and atmospheric sciences. Unlike other books that focus on this topic, this book provides questions at the end of each chapter that encourage students to grapple with the material, both conceptually and quantitatively. Even better, there is explicit connection to the peer-reviewed literature that encourages students to see applications of the concepts to the practice of science. The connection of the material to observational data sets through the *Dive into the Data* boxes in each chapter introduces students to data-driven discovery in ocean sciences.'
– LuAnne Thompson, University of Washington

'A new textbook that captures the physics of our coupled atmosphere-ocean system. This is no ordinary textbook. It takes us on a journey in exploring and understanding the physics of our planet's two fluids (one ocean, one atmosphere) and how they talk to each other. Using an easy style of writing without compromising the mathematics, Karnauskas draws us into some of the history of scientific discovery of our ocean and atmosphere, while also presenting the physics and mathematics of fluid dynamics and the interactions between these two fluids. Each chapter poses questions for furthering the reader's discovery and thinking, often through the use of data sets. The book leads us ultimately to a discussion on the atmosphere-ocean system by describing climate not as an average of weather, but as an interacting system that produces climate variability and change. I highly recommend this textbook, written by a new leader in the study of our planet.'
– Susan K. Avery, Woods Hole Oceanographic Institution

' ... an important and timely text that focuses squarely on the role of the ocean in the climate system. It is cleverly organized to present the oceanography that will help the reader understand the role of the ocean in climate dynamics. It is written in an accessible form that make it valuable both as a textbook and a reference book.'
– Enrique Curchitser, Rutgers University

'*Physical Oceanography and Climate* fills a gap in the scientific literature at a time when the societal issues surrounding climate variability and change are becoming ever more urgent. This engagingly written book, with its focus on fundamentals and hands-on learning exercises, is a great introduction to the ocean's role in climate and why it matters. It will serve as a valuable resource for students and research scientists interested in the processes that govern ocean-atmosphere interactions and their consequences for the climate system.'
– Michael J. McPhaden, University of Washington

Physical Oceanography and Climate

Kris Karnauskas

University of Colorado Boulder

CAMBRIDGE
UNIVERSITY PRESS

CAMBRIDGE
UNIVERSITY PRESS

Shaftesbury Road, Cambridge CB2 8EA, United Kingdom

One Liberty Plaza, 20th Floor, New York, NY 10006, USA

477 Williamstown Road, Port Melbourne, VIC 3207, Australia

314–321, 3rd Floor, Plot 3, Splendor Forum, Jasola District Centre, New Delhi – 110025, India

103 Penang Road, #05–06/07, Visioncrest Commercial, Singapore 238467

Cambridge University Press is part of Cambridge University Press & Assessment, a department of the University of Cambridge.

We share the University's mission to contribute to society through the pursuit of education, learning and research at the highest international levels of excellence.

www.cambridge.org
Information on this title: www.cambridge.org/9781108423861

DOI: 10.1017/9781108529594

First published 2020

A catalogue record for this publication is available from the British Library

Library of Congress Cataloging-in-Publication data
Names: Karnauskas, Kris, author.
Title: Physical oceanography and climate / Kris Karnauskas.
Description: Cambridge, United Kingdom ; New York, NY : CambridgeUniversity Press, 2020. | Includes bibliographical references and index.
Identifiers: LCCN 2019047036 (print) | LCCN 2019047037 (ebook) | ISBN 9781108423861 (hardback) | ISBN 9781108529594 (ebook)
Subjects: LCSH: Ocean-atmosphere interaction. | Climatic changes.
Classification: LCC GC190.2 .K37 2020 (print) | LCC GC190.2 (ebook) | DDC 551.46–dc23
LC record available at https://lccn.loc.gov/2019047036
LC ebook record available at https://lccn.loc.gov/2019047037

ISBN   978-1-108-42386-1   Hardback

Brief Contents

The plate section can be found between pp. 132 and 133.

Contents

The plate section can be found between pp. 132 and 133.

Dive into the Data

Written by a prolific research scientist, *Physical Oceanography and Climate* makes use of dozens of publicly available, research-quality data sets to render vivid illustrations of the core scientific concepts taught throughout the book. *Dive into the Data* is a book feature that enables students (and instructors) to actually *use* real, research-quality data sets to go beyond the graphs. There is one *Dive into the Data* box for every major data set used – roughly three per chapter, or 27 in total. These boxes are meant to accelerate the process of hands-on learning and exploration regardless of skill level; they can form the basis for a wide range of homework assignments, lab activities, term projects, etc.

Each *Dive into the Data* box includes the full data set name and version, lists which figures in the book utilize the data set, some key metadata such as the spatial and temporal aspects of the data set (domain and resolution), a plain-language synopsis of the data set and its typical scientific uses, the original source (URL) of the data set and what format is found there, and the filenames of the value-added resources provided by the book author (neatly packaged, easy-to-use MATLAB data files and sample codes) available at www.cambridge.org/karnauskas. Finally, the questions at the end of each chapter include a subset of questions that encourage (or require) students to access and analyze the data sets described in *Dive into the Data* boxes. Following is a complete list of *Dive into the Data* boxes by chapter.

Chapter 1

1. Antarctic Composite Ice Core Atmospheric CO_2
2. Atmospheric CO_2 at Mauna Loa Observatory
3. Representative Concentration Pathway (RCP) Global Mixing Ratios of Atmospheric CO_2

Chapter 2

1. WHOI Ocean–Atmosphere Flux (OAFlux)
2. NOAA Optimal Interpolation (OI) Sea Surface Temperature (SST)
3. NOAA Tropical Atmosphere–Ocean (TAO) Array

Preface

This book was written based on the need for an accessible yet quantitative treatment of physical oceanography for those students whose true motivation is to understand climate variability and change. It was born out of a course by the same name, which I developed and have taught in the Department of Atmospheric and Oceanic Sciences at the University of Colorado Boulder since the fall of 2015. There are quite a few programs like ours nowadays, where oceanography and atmospheric science fall under a single undergraduate curriculum. Moreover, the broad recognition across the Earth sciences of the importance of interdisciplinary competency has brought many graduate students to my classroom from closely related disciplines ranging from paleoclimatology to science journalism. This tells me there is a steadily growing population of students who are studying other realms of the climate system (atmosphere, cryosphere, biosphere, carbon cycle, etc.), not to mention an informed public, who want to know more about how the ocean fits in. That is exactly the role this book aims to fill.

To that end, this is not a traditional "GFD" (geophysical fluid dynamics) book. Some topics that one might expect to see in such a pure physical oceanography text, like surface waves or tides, are omitted here in the interest of keeping the material germane to those motivated primarily by climate. This book provides a course on global climate dynamics from the perspective of the upper ocean – the mixed layer, to be specific. Why do sea surface temperature anomalies develop? How does ocean salinity respond to the atmosphere like a rain gauge? What drives the circulation of the ocean, and where are the parallels with the governing equations of the atmosphere? This book introduces these fundamentals in a unified budget framework that quickly becomes familiar to the student as we apply it to the conservation of energy, mass, and momentum. Those budgets are cast in such a way as to deliberately and readily identify the points of contact between the ocean and atmosphere, leading to the wind-driven and thermohaline circulations, the mechanisms by which the resultant variations in temperature at the ocean surface influence the atmospheric circulation, and how those interactions give rise to *coupled* climate variability from interannual to multidecadal and longer time scales. Finally, the closing chapter offers a grand view of anthropogenic radiative forcing and global climate models before delving into some of the pathways by which climate change rears its head in the ocean.

The book is ideal for a one-semester, upper-level undergraduate course in a department of atmospheric science and/or oceanography, or an elective graduate course in virtually any department within the Earth sciences. At the University of Colorado Boulder it is offered as both at once, and that works great, too. I typically divide the semester into three units with an exam following each. After setting the stage, the first unit introduces conservative budgets in the form of partial differential equations to explore what sets the temperature and salinity of the upper ocean (Chapters 1–3). The second unit delves into dynamics, culminating in an understanding of the wind-driven ocean circulation (Chapters 4–6). The final unit focuses on variability and change in the climate system with an emphasis on ocean-atmosphere coupling (Chapters 7–9). Alternatively, one could split the course into two halves: one covering all of the underlying budgets through momentum (Chapters 1–4), and one rich in applications to steady circulations and climate variability alike (Chapters 5–9).

The writing in this book closely reflects my own teaching style. I'm precise, but not always too formal. The dynamics and mathematical derivations are accompanied by plain-language discussion and schematics. This is a skill possessed by some of my own teachers who I strive to emulate – Jonathan Martin at the University of Wisconsin Madison comes to mind as a master of the craft. I have also made a deliberate attempt in each chapter to highlight the diversity of ocean and climate scientists making exciting progress and discoveries in this field. My years of engagement with US CLIVAR serve as one important source of inspiration in that regard. I am always inspired by my friends, peers, colleagues and mentors, and I think they all deserve to have their work highlighted in textbooks being used to educate the next generation.

I would like to give a special thanks to Matt Lloyd and the entire team at Cambridge University Press for their patience and guidance through the process of developing and writing a book. I thank Elizabeth Maroon, Raghu Murtugudde, Ray Schmitt, and other anonymous reviewers who provided helpful comments on draft chapters. A very hearty thanks also goes out to one of the brightest undergraduates the University of Colorado Boulder has ever seen, Michelle Maclennan, for a thorough and insightful review of the complete work from the most important perspective of all: that of the student. This book and its many illustrations draw from an extensive wealth of observational and model data sets. Sources are always acknowledged in the captions, but I would like to express my gratitude for all of the observational teams, climate modeling groups (e.g., CMIP5) and government science agencies who ensure that such data sets are made freely accessible to the public.

Finally, I could not have written this book without the patience, love, and support of my family, to whom this book is dedicated: Alexis, Dean, and Caroline Karnauskas.

Introduction

"How inappropriate to call this planet Earth when clearly it is Ocean."

Arthur C. Clarke (Lovelock, 1990)

1.1 From Blue Marbles to Pale Blue Dots

Human curiosity and ingenuity have given rise to some truly remarkable ways to look back at our home planet. It is the twenty-first century, and many of us have the ability to view planet Earth from just about any distance, angle, zoom factor, or map layer (including the seafloor) using a device that fits in a pocket. Sure, scholars have suspected planet Earth is round for thousands of years and have long been aware of the fragile dimensions of our atmosphere and ocean compared to the massive spherical rock to which they cling, but all without actually *seeing* it. You may be surprised just how recently we acquired such a basic perspective on our home planet.

A short history of planetary selfies takes us to March 1946, on Santa Monica Boulevard in Hollywood, California. Modern civilization was in full swing. Two World Wars fully behind us, Charlie Parker and Miles Davis (Figure 1.1) were recording the eventual Grammy Hall of Fame title *Ornithology*, and yet we had no idea what Earth looked like from anywhere but Earth! That started to change later that year, just a thousand kilometers due east in the desert of New Mexico, where a group of rocket scientists sent a heisted German V-2 rocket to altitudes where the line is blurred between atmosphere and outer space. This was the first photograph of our home planet taken from "space" (105 km), using a black and white 35 mm camera and dropped back to the sand in a rugged tin can (Figure 1.2). It wasn't much by twenty-first-century standards, but what a profound sight for humanity and a society so endlessly preoccupied with our own affairs. This snapshot gave our atmosphere a three-dimensional character; one can perceive space between the clouds and their shadows on the ground. One may also appreciate from this vantage point how little regard the climate has for sovereign borders. Cloudy skies were finite, and even oceans that seemed impossibly far away weren't so far at all.

Figure 1.1 [Year 1947, distance 0 km] Photo taken on the surface of Earth by William Gottlieb of Charlie Parker and Miles Davis playing at the Three Deuces jazz club in New York ca. August 1947. The first photo *of* Earth (from space) was taken less than one year earlier. Credit: William P. Gottlieb/Ira and Leonore S. Gershwin Fund Collection, Music Division, Library of Congress.

Figure 1.2 [Year: 1946, distance: 105 km] The first photo of Earth from space, taken from aboard a German V-2 rocket launched from White Sands Missile Range in New Mexico on October 24, 1946. Credit: U.S. Army White Sands Missile Range/Johns Hopkins Applied Physics Laboratory.

Since that first photograph of Earth from space in October 1946, humans only continued to go farther up, but never stopped looking back. John Glenn snapped a color photo in 1962 while orbiting Earth; even 266 km wasn't far enough away to frame the whole planet in a single shot, but both the roundness of Earth and the dominance of water were vividly clear (Figure 1.3). It wasn't until the late 1960s when the "whole" Earth was captured on film, ironically because of how irresistible a visit to the barren lunar surface was. Two iconic photos of Earth were captured by separate Apollo crews en route to the Moon: *Earthrise* in 1968 and the *Blue Marble* in 1972 (Figure 1.4). Perhaps the most profound photographic perspective on planet Earth that humanity ever acquired came from a distance so great it could hardly be seen at all. Just as the Voyager 1 spacecraft was leaving our solar system, astronomer Carl Sagan convinced NASA administrators to point Voyager's camera back toward the inner solar system and take one, final photo of home

Figure 1.3 [Year: 1962, distance: 266 km] Photo of Earth taken by astronaut John Glenn on February 20, 1962 aboard the Mercury spacecraft during his flight as the first American to orbit the Earth. Credit: NASA/John Glenn. (A black and white version of this figure will appear in some formats. For the color version, please refer to the plate section.)

Figure 1.4 [Year: 1972, distance: 29,000 km] Photo of Earth, dubbed "Blue Marble," taken by the crew of Apollo 17 on December 7, 1972 while en route to the Moon. Photo credit: NASA/Apollo 17 crew. (A black and white version of this figure will appear in some formats. For the color version, please refer to the plate section.)

Figure 1.5 [Year: 1990, distance: 5,954,572,800 km] Photo of Earth, dubbed "Pale Blue Dot," taken by the Voyager 1 space probe upon leaving the Solar System on February 14, 1990. Credit: NASA/Voyager 1. (A black and white version of this figure will appear in some formats. For the color version, please refer to the plate section.)

(Figure 1.5). The resulting image, taken in 1990, inspired deep contemplation about our place in the universe. As Sagan himself offered: "There is nowhere else, at least in the near future, to which our species could migrate. Visit, yes. Settle, not yet. Like it or not, for the moment the Earth is where we make our stand" (Sagan, 2004).

Each successive photograph of Earth, down to the naming of the infamous Voyager 1 shot – the *Pale Blue Dot* – further illuminated the significance of the ocean to humanity. We don't live in it, yet we cannot live without it. The ocean covers about 70 percent of the planet; one can even spin a globe (or Google Earth) to an angle from which hardly a speck of land can be seen. Likewise, we've also come to understand what a vital role the ocean plays in global climatology, which is exactly the motivation of this book. There are so many wonderful and important books on descriptive physical oceanography and geophysical fluid dynamics, but this book adopts a perspective on the ocean's physics from the very point where it interacts with the atmosphere vis-à-vis climate dynamics. How does the ocean work, and how does it fit into climate and Earth system science writ large? This is now, and will continue to be, a crucial field of study to be applied to the management, sustainability, and continued habitability of the global environment.

From the first black and white TIROS images in 1960 to state-of-the-art satellite altimeters measuring global sea level variations with centimeter precision, to the very latest profiling float dropped into the sea (of which there are nearly 4000 active at a given time), we are probing Earth's dynamic climate system, including the ocean, continuously and from every angle. Much like a doctor examining a patient, this endeavor to "look back" at our home planet in such a multitude of ways has enabled us to carefully monitor its internal rhythms as well as detect and diagnose longer-term changes. Many of the variations to be examined, as we shall throughout this book, are perfectly natural and are expressions of the rich dynamical cooperation between the ocean and atmosphere. Unfortunately, some of these variations are not natural, and are cause for grave concern about the overall health of the global environment.

Sustained measurements of atmospheric carbon dioxide (CO_2) taken from high above the North Pacific Ocean since 1958 have revealed an exponentially growing concentration – an increase by about 93 ppm as of 2018. What is remarkable about this trend is not only the amount – it is roughly the amount by which CO_2 varied across the ice age cycles of the Pleistocene epoch – but the *pace* at which it is rising. Human activities, in particular fossil fuel combustion, are increasing CO_2 concentrations faster than at any time in at least the past million years, and we may be on course to reach nearly 1000 ppm by the end of the twenty-first century (Figure 1.6). The ocean plays a key role in mediating the response to this forcing for the entire climate system, including the atmosphere, *and* delivers some of the major impacts directly. The ocean has already absorbed some 30 percent of the anthropogenic CO_2 emitted, which is a double-edged sword at the front lines of climate

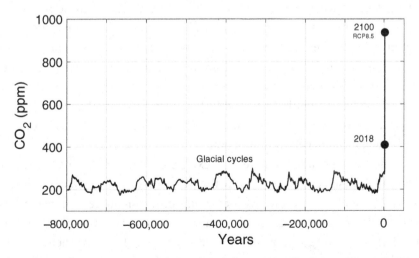

Figure 1.6 Atmospheric concentration of carbon dioxide (parts per million) since about 800,000 years ago from Antarctic ice core records, direct measurements at Mauna Loa Observatory, and future predictions based on unmitigated fossil fuel emissions.

change. On one hand, that's 30 percent less CO_2 remaining in the atmosphere, which at least temporarily dampens the level of greenhouse forcing. On the other hand, the CO_2 entering the ocean has severe consequences for marine life (e.g., through ocean acidification). Thus, physical oceanography acts as something of a mediator between climatic change and the other various subdisciplines of oceanography, including marine biology and chemical (or biogeochemical) oceanography. Meanwhile, the ocean is also absorbing the majority of the excess heat trapped near the surface by greenhouse gases. Due to the great heat capacity of water, the ocean has served to slow the overall warming of the climate system as we humans feel it, but the upper ocean has indeed warmed and expanded. The more visceral consequences of climate change, such as rising seas and worsening tropical storms, are striking but still remain dependent on how much CO_2 we decide to emit over the coming decades. Central to understanding just about any flavor of climate variability, from El Niño to global warming and beyond, are the physical laws governing the exchanges of energy between the ocean and atmosphere, and how these exchanges ultimately control the circulation of both fluids. These chapters aim for such a literacy of ocean physics for students across the Earth and climate sciences.

1.2   Climate is a Coupled System

The last several decades have seen a steady convergence of two fields: atmospheric science and oceanography. Today, you are more likely to find a university where these two disciplines have been deliberately blended into a single department (e.g., Atmospheric and Oceanic Science), rather than historically as standalone departments. It was not always recognized just how closely related these two fields are, nor how important it is to foster cross-disciplinary research to solve the many riddles of climate. Atmospheric scientists now appreciate that the ocean is more than just a two-dimensional boundary condition, and physical oceanographers have learned how difficult it is to predict the ocean currents without accounting for the evolution of the wind field both near and far. In fact, some of the most important climatic phenomena are impossible to describe, let alone predict, without invoking *coupling* between the ocean and atmosphere.

There is perhaps no more famous an example of this coupling, evidenced by its awkwardly constructed name, than the El Niño–Southern Oscillation (ENSO). In what *must* have been a strictly oceanographic phenomenon, the occasional warming of waters off the coast of Peru in the eastern tropical Pacific around Christmastime was long known as El Niño. Historically, the science around El Niño involved ocean temperature, currents, etc., but only *local* weather impacts (driven by the warmer-than-usual ocean). Meanwhile, on the other side of the world, Sir Gilbert Walker was poring over meteorological records at the Indian Meteorological Department and discovered a vast seesaw pattern in atmospheric pressures across the Indian and Pacific Oceans whose strength also varied from year to year (then termed the Southern Oscillation) (Walker, 1923). It wasn't until the late 1960s that meteorologist Jacob Bjerknes successfully argued that the variations in the Walker cell and the occasional warmings in the eastern equatorial Pacific Ocean were linked, hence the El Niño–Southern Oscillation (Bjerknes, 1969). It is quite fitting that Bjerknes' (and his father's) scientific career was dedicated to forecasting, because it was only after this recognition set in, and detailed study by teams of oceanographers *and* atmospheric scientists such as the Tropical Ocean–Global Atmosphere (TOGA) program of the 1980s and 1990s (McPhaden *et al.*, 1998; Figure 1.7), that ENSO could be predicted with any skill and the connections to seasonal weather anomalies around the globe were harnessed.

The implications of ocean–atmosphere coupling have global reach and extend throughout the depths of both fluids. At a basic level, though, the ocean and atmosphere achieve their coupling locally, and right at their physical interface – the surface. The ocean directly feels the atmosphere above through a multitude of physical processes, which can generally be categorized as fluxes of heat, freshwater, and momentum. We will break down the heat fluxes in the next chapter. Freshwater flux results from a local imbalance between precipitation and evaporation (Chapter 3) and helps determine the density, and thus the buoyancy, of water near the surface – a major driver of the global ocean circulation that we will learn more about in

Figure 1.7 Photo of a mooring that simultaneously probes both the upper ocean and lower atmosphere, part of the Tropical Atmosphere–Ocean (TAO) array in the tropical Pacific Ocean – an important legacy of the TOGA program. Photo credit: Kris Karnauskas.

Chapter 8. Momentum flux is imparted locally onto the ocean surface from the wind (Chapters 4 and 5) and is propagated downward into the ocean's interior by friction; the spatial patterns of wind stress turn out to be very important in driving the upper ocean circulation (Chapter 6). In turn, the atmosphere is primarily influenced by the ocean through the exchange of heat and emission of radiant energy. For example, a warmer ocean surface emits more longwave radiation; evaporation and thus latent heat flux is also likely to be greater. Just how the atmosphere as a whole reacts to surface fluxes will also be examined in greater detail in Chapter 5, but it is abundantly clear that the equilibrium atmosphere is directly constrained by fluxes at the ocean surface. Particularly in the tropics, deep rising atmospheric motion and thus rainfall is all but a perfect reflection of warm sea surface temperatures (Figure 1.8). Come two-thirds of the way into this book, you will be able to look at either map in this figure and explain why the salient features look just so, and elaborate on the role of ocean–atmosphere interaction in setting those patterns. These are the fundamentals; such is how we will treat the system at first, before returning to what happens when we consider that – like the synergy between Charlie Parker and Miles Davis in live performance – all of these processes are

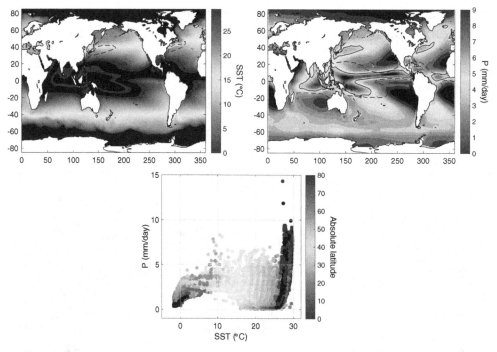

Figure 1.8 Maps of time-averaged sea surface temperature (SST; °C) and precipitation (mm/day) over the global ocean (top row), and their correspondence colored by the absolute value of latitude (bottom panel). On the SST map, the 27 °C and 5 mm/day contours are shown in solid and dashed, respectively, and vice versa on the precipitation map. SST and precipitation observations from the NOAA OIv2 and GPCP data sets, respectively; both averaged from 1982 through 2018. (A black and white version of this figure will appear in some formats. For the color version, please refer to the plate section.)

interactive and the actual state of the climate system is never in equilibrium for long, making way for feedbacks and a beautiful spectrum of climate variability.

1.3   Our Common Framework

As articulated above, the time has never been better to understand Earth's climate system and how it evolves. This is essentially a book about climate dynamics, but from the unique perspective of the upper ocean. To that end, a little math will go a long way. Three mathematical constructs permeate this book: **budgets**, **vectors**, and **partial derivatives**. When combined, they facilitate a deeper understanding of the workings of the ocean and its role in the climate system than words alone might. Even the student less familiar with these tools needn't worry; let's take a crash course.

1.3.1   The Budget Paradigm

The next three chapters build an understanding of what controls the heat, salt, and momentum of the upper ocean using a consistent budget framework. Such budgets are nothing more than usable, mathematical expressions of conservation laws. They are partial differential equations rearranged into **Eulerian** statements of what matters and how. Eulerian means taking the perspective of a fixed point within the fluid, rather than following the fluid (which would be a "Lagrangian" perspective). Each of our budgets have a similar structure that will quickly become familiar. The **terms**, or groups of variables, in our budget equations usually represent external influences on the budgets in question, plus processes internal to the fluid itself (such as ocean currents moving heat around, or swirling eddies mixing salty and freshwater together). Such terms in each budget have physical meaning that you will very soon recognize and think about like second nature. They will then serve as the basis for all of the subsequent chapters; fortunately, we will mostly be scaling them down in an attempt to explain the ocean and climate with as few unnecessary complications as possible.

1.3.2   Vectors

Our budgets will contain some vector expressions, because the ocean variables that these budgets aim to diagnose (temperature, salinity, and currents) can be affected by the three-dimensional currents within the ocean. Vectors and vector calculus are deep subjects that years of mathematical coursework can be dedicated to. Here is what you need to know about vectors in order to understand the material in this book. We live in a three-dimensional world (as in the Cartesian dimensions x, y, and z), which means seawater at any particular location might be moving a little bit eastward, southward, too, and sinking all at the same time. Constituting both a speed and direction, vectors are useful for describing fluid velocity. Specifically, anything in motion like wind or an ocean current should be described by a combination of how rapidly it is flowing in the x direction (i.e., eastward or westward), which we'll assign to the variable u (all scalar variables will be italicized throughout this book), how much it is flowing in the y direction (i.e., northward or southward), which we'll call v, and how much it is flowing in the z direction (i.e., upward or downward), which we'll call w (Table 1.1). These velocity components u, v, and w are just the usual derivatives of position with respect to time (dx/dt, dy/dt and dz/dt) and so will each have units of meters per second (m/s). Earth scientists – especially atmospheric scientists and oceanographers – like to refer to the x dimension as **zonal** and the y dimension as **meridional** (while "vertical" is just fine for the z dimension), so u is referred to as the zonal component of velocity (or just "zonal velocity"), and so on. Finally, although we generally neglect these from our notation, it is worth a reminder that unit vectors $\hat{\imath}$, $\hat{\jmath}$, and $\hat{k}$ (pronounced "i-hat," and so on) point in the zonal, meridional, and vertical directions, respectively, with a magnitude of 1 m/s.

So, when one wishes to describe the velocity of a fluid like seawater at some particular time and geographic location, one may write it as $u\,\hat{\imath} + v\,\hat{\jmath} + w\,\hat{k}$, or simply **V** for short (vectors will be bolded throughout this book). This is a good time to bring up one of the *operations* out of the vector calculus playbook that we'll use from time to time: the dot product, which is rather simple – in our application of it, anyway. To take the dot product (·) of two vectors, say $(2\,\hat{\imath} + 3\,\hat{\jmath} + 5\,\hat{k}) \cdot (1\,\hat{\imath} + 4\,\hat{\jmath} - 2\,\hat{k})$, we simply multiply the zonal, meridional, and vertical *components* from the two vectors with one another, yielding $2\,\hat{\imath} + 12\,\hat{\jmath} - 10\,\hat{k}$. This will take on physical significance when, for example, we "dot" velocity vectors with spatial gradients (which are composed of partial derivatives) of seawater properties such as temperature to define an important process known as advection.

1.3.3   Partial Derivatives

Did you forget what a partial derivative is? No problem! Even if you have taken years of partial differential equation (PDE) courses, you may not be accustomed to applying them in an Earthly context and so you, too, could use a little practice evaluating them by eye. If you recall that a derivative is the rate of change of some

Table 1.1 Symbols used for velocity components and unit vectors for the three Cartesian dimensions, and the names commonly used in physical oceanography and climate science to refer to them.

Cartesian dimension	Velocity component	Unit vector	Climate jargon
x	u	$\hat{\imath}$	Zonal
y	v	$\hat{\jmath}$	Meridional
z	w	$\hat{k}$	Vertical

variable (with respect to others), then partial derivatives are a breeze – again, especially as applied throughout this book. Partial differentiation is about acknowledging that we live in a four-dimensional world (including time), but only thinking about one dimension at a time. Specifically, the qualifier "partial" simply means the rate of change of a variable *with respect to only one dimension* (holding all others constant), and *at just one place and time*. When we apply these constraints, we express the derivative as a partial one by using the symbol ∂ (rather than d or D).

Consider, for example, a variable like temperature T that happens to be a function of all three spatial dimensions (x, y, z) as well as time (t), which we thus express that as $T(x, y, z, t)$. As you might imagine, this is typically the case in the ocean, as well as the atmosphere and many other fluid media. The tropical oceans are relatively warm, deep water is relatively cold, and ocean temperatures change from day to day, year to year, and so on. We can depict our variable T in three separate perspectives: a **time series** $T(t)$ at location x_0, y_0 and depth z_0, a map or **field** $T(x, y)$ at depth z_0 and time t_0, and a **profile** $T(z)$ at location x_0, y_0 and time t_0 (Figure 1.9). Using asterisks (*),

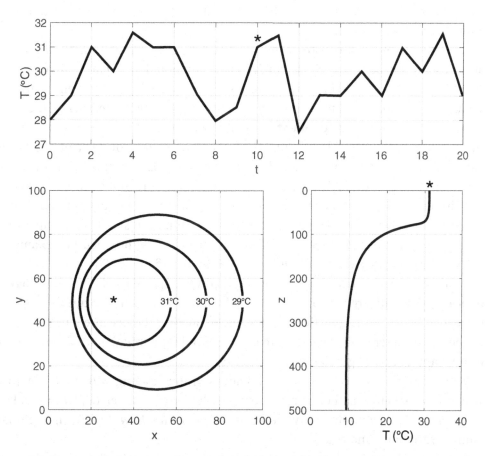

Figure 1.9 Three graphical representations of a hypothetical variable $T(x, y, z, t)$: $T(t)$ at x_0, y_0, z_0, $T(x, y)$ at z_0 and t_0, and $T(z)$ at x_0, y_0 and t_0. Asterisks in each panel denote x_0, y_0, z_0, and t_0. Use the illustrations to practice estimating partial derivatives of T.

the specified depth z_0 is shown on the profile (at $z = 0$, or the surface), the specified location x_0, y_0 is marked on the map (at $x = 30$, $y = 50$), and the specified time t_0 is marked on the time series (at $t = 10$). Let's start with the map of $T(x, y, z_0, t_0)$, and eyeball some partial derivatives. If we focus on the area around, say, $x = 80$, $y = 50$, what partial derivatives *can* we evaluate? We are given no information on how T is changing with time or depth at that particular location, but we can see how T is (or is not) changing with respect to x and y there. Scanning from left (i.e., from slightly lesser x values than 80) to right (i.e., toward values greater than 80 along the x axis), at $y = 50$, T is clearly decreasing, so we conclude that the partial derivative of T with respect to x, or $\partial T/\partial x$, at that location, depth, and time is negative. If the horizontal coordinates x and y are in meters, then the zonal temperature **gradient** $\partial T/\partial x$ there has a value of approximately -1 °C per 20 m or -0.05 °C/m. If we try to estimate the *meridional* temperature gradient ($\partial T/\partial y$) at the same location, we conclude that it must be zero, since the isotherms (lines of constant temperature) run parallel to the y axis at location $x = 80$, $y = 50$. Now ask yourself how the zonal temperature gradient $\partial T/\partial x$ on the other side of the hot spot (say, $x = 15$, $y = 50$) compares to the one at our previous location, in terms of both sign and magnitude.

A final mathematical construct that we should briefly review in light of its frequent use in this book, which is actually just a combination of partial derivatives, is the del operator (∇). When applied to a scalar field such as, say, temperature T, salinity S, or zonal velocity u, the del operator ∇ is formally defined as the full, three-dimensional spatial **gradient** of that scalar field: $\partial/\partial x\ \hat{\imath} + \partial/\partial y\ \hat{\jmath} + \partial/\partial z\ \hat{k}$. Do you notice the variable missing from the "numerator" of each of the three partial derivatives? Simply insert the scalar field whose gradient you are considering, so $\nabla T = \partial T/\partial x\ \hat{\imath} + \partial T/\partial y\ \hat{\jmath} + \partial T/\partial z\ \hat{k}$. As a reminder, we will avoid the clutter of unit vectors in the mathematical notation throughout this book. When the del operator is applied to a vector, it cannot be simply its gradient as with scalars; it can, however, be through a dot (or cross) product. For example, $\nabla \cdot \mathbf{V} = (\partial/\partial x\ \hat{\imath} + \partial/\partial y\ \hat{\jmath} + \partial/\partial z\ \hat{k}) \cdot (u\ \hat{\imath} + v\ \hat{\jmath} + w\ \hat{k}) = \partial u/\partial x\ \hat{\imath} + \partial v/\partial y\ \hat{\jmath} + \partial w/\partial z\ \hat{k}$. As we will see in later chapters, the dot product $\nabla \cdot \mathbf{V}$ represents the velocity **divergence**. Both the divergence of the velocity field and its cousin the **curl** ($\nabla \times \mathbf{V}$) are extremely powerful diagnostic tools for understanding the ocean circulation. See the end-of-chapter problems for more practice thinking through partial derivatives (spatial and temporal gradients) as necessary.

With some relevant context and motivation in mind, along with some comfort with a few essential mathematical constructs, let's try to understand what makes the ocean tick, how it interacts with the atmosphere, and how it all fits into global climate variability and change.

Further Reading

IPCC (2013) is the go-to synthesis on historical and future climate change.

Questions

1. Translate the following mathematical expression into climate jargon: $\partial v/\partial x$.

2. Estimate $\partial CO_2/\partial t$ at some time during the second-to-last deglaciation (roughly 130,000 years ago), and at the year 2000 CE. For a quantitative approach, you may wish to utilize *Dive into the Data* Boxes 1.1 and 1.2.

3. Estimate the zonal temperature gradient $\partial T/\partial x$ to the left of the hot spot in Figure 1.9 ($x = 15$, $y = 50$). How does it compare to that on the right side of the hot spot ($x = 80$, $y = 50$) both in terms of sign and magnitude? Assume the x and y axes are in units of meters.

4. Flip forward to Figure 5.1. Reproduce (by sketching) the graph shown. Decide how to divide up the temperature profile into a handful of layers. Label them A, B, C, and so on. Rank your layers of the atmosphere and ocean according to their vertical temperature gradients $\partial T/\partial z$, from strongest negative to strongest positive. Define positive z as upward.

5. Which photograph of Earth displayed in this chapter gives you the most stirring impression? What observation do you find interesting about it?

6. If you are a newcomer to climate science and/or oceanography, is there a specific variable or quantity you think we should be (or hope that we are) observing routinely by some means other than a photographic camera? Why do you think that variable might be important to monitor? How do you think such a sustained Earth observation might be achieved?

7. Select one of the *Dive into the Data* boxes featured in this chapter. Read the sample journal article that uses the associated data set, and describe the general role of that data set in the study. Provide a little context, both technical and scientific: What did the authors actually "do" with the data set? What specific scientific insight was enabled by incorporating this data set into the study?

8. There are about four decades of overlap between the Antarctic composite CO_2 record and the Mauna Loa CO_2 record, and 14 years of overlap between the Mauna Loa CO_2 record and the Representative Concentration Pathway (RCP) CO_2 time series. Utilizing *Dive into the Data* Boxes 1.1, 1.2, and 1.3, comment on how well the Antarctic and Mauna Loa records match, and comment on your comparison between the Mauna Loa record and the four RCP time series.

DIVE INTO THE DATA BOX 1.1

Name   Antarctic Composite Ice Core Atmospheric CO_2

Synopsis   The Antarctic Composite Ice Core Atmospheric CO_2 data set is a *composite* of about 10 different CO_2 records covering various time periods, all extracted from the Antarctic ice sheet. Some records overlap in time with one another, in which case they are averaged. CO_2 records are extracted from ice sheets by "coring" miles deep into them and extracting cores that are ~30 m long and ~10 cm across. The air bubbles trapped in these ice cores act like a time capsule of the atmospheric concentration of CO_2 (and other constituents), with deeper into the ice representing earlier in time. Since CO_2 is a well-mixed gas, the Antarctic composite record is usually interpreted to represent global atmospheric CO_2 concentration.

Science   Notable ice cores shedding significant light on climate changes over the past 800,000 years have been retrieved from Earth's major ice sheets, including Antarctica (this data set) and Greenland. The Greenland ice core records are shorter than those of Antarctica but yield information with finer temporal resolution because of the higher rate of snow and ice accumulation on Greenland. Thus, the Antarctic Composite record provides a view of the ice age cycles of the past 800,000 years, while Greenland gives finer detail on how the most recent glacial–interglacial cycle transpired. The Antarctic CO_2 record is how we know that prior to the Industrial Revolution, atmospheric CO_2 concentration bounced between about 200 and 300 ppm, reaching a lowest value of 172 ppm seven ice ages ago and a peak of 300 ppm three interglacials ago, in stark contrast to the rise to over 410 ppm in just the past century.

Figures   1.6, 9.1

Version   Revised 2015

Variable   Atmospheric concentration of carbon dioxide

Platform   Ice core

Spatial   1D (time series), Antarctica

Temporal   803,719 BCE to 2001 CE, variable spacing but on average a few hundred years

Source   http://ncdc.noaa.gov/paleo/study/17975

Format   Excel (.xls)

Resource   co2.m; co2.mat (sample data and code provided on publisher website here: www.cambridge.org/karnauskas)

DIVE INTO THE DATA BOX 1.1 (cont.)

Journal References for Data

Lüthi, D., Le Floch, M., Bereiter, B., *et al.* High-resolution carbon dioxide concentration record 650,000–800,000 years before present. *Nature* **453**, 379–382 (2008). DOI: 10.1038/nature06949.

Bereiter, B., Eggleston, S., Schmitt, J., *et al.* Revision of the EPICA Dome C CO_2 record from 800 to 600 kyr before present. *Geophys. Res. Lett.* **42**, 542–549 (2015). DOI: 10.1002/2014GL061957.

Sample Journal Article Using Data

Cheng, H., Lawrence Edwards, R., Sinha, A., *et al.* The Asian monsoon over the past 640,000 years and ice age terminations. *Nature* **534**, 640–646 (2016). DOI: 10.1038/nature18591.

DIVE INTO THE DATA BOX 1.2

Name	Atmospheric CO_2 at Mauna Loa Observatory
Synopsis	Direct chemical measurements of atmospheric CO_2 concentration at Mauna Loa Observatory, on the northern flank of Mauna Loa Volcano, Big Island, Hawaii. Since CO_2 is a well-mixed gas and Mauna Loa Observatory is far from major emission sources, the Mauna Loa record is interpreted to represent global atmospheric CO_2 concentration, although some offset with the true global average concentration will exist since Mauna Loa Observatory is at an altitude of 3397 m.
Science	The Mauna Loa record, also known as the "Keeling Curve," reveals how the atmospheric CO_2 concentration has changed over the past several decades as human emissions (e.g., fossil fuel burning) progress. The measurements have revealed an increase in global atmospheric CO_2 concentration from 316 ppm in 1959 to 408.5 ppm in 2018 (29 percent). Natural rhythms of the planet are also documented in the Mauna Loa records, such as the effect of El Niño on atmospheric CO_2 concentration and the seasonal cycle – due to the hemispherically asymmetric distribution of continents (more land in the Northern Hemisphere) and thus photosynthesizing flora.
Figures	1.6, 9.1
Version	N/A
Variable	Atmospheric concentration of carbon dioxide

DIVE INTO THE DATA BOX 1.2 (cont.)

Platform   Nondispersive infrared analyzer
Spatial   1D (time series), Hawaii
Temporal   March 1958 to present, hourly to annual averages
Source   www.esrl.noaa.gov/gmd/ccgg/trends/data.html
Format   Plain text (.txt)
Resource   co2.m; co2.mat (sample data and code provided on publisher
               website here: www.cambridge.org/karnauskas)

Journal Reference for Data

Keeling, C. D., Bacastow, R. B., Bainbridge, A. E., *et al.* Atmospheric carbon
    dioxide variations at Mauna Loa Observatory, Hawaii. *Tellus* **28**, 538–551
    (1976). DOI: 10.1111/j.2153-3490.1976.tb00701.x

Sample Journal Article Using Data

Bacastow, R. B., Keeling, C. D., and Whorf, T. P. Seasonal amplitude increase in
    atmospheric CO_2 concentration at Mauna Loa, Hawaii, 1959–1982. *J. Geophys.
    Res.* **90**(D6), 10,529–10,540 (1985). DOI: 10.1029/JD090iD06p10529.

DIVE INTO THE DATA BOX 1.3

Name   Representative Concentration Pathway (RCP) Global Mixing Ratios
               of Atmospheric CO_2
Synopsis   Set of four possible future *pathways* of atmospheric CO_2
               concentrations, depending on various assumptions about drivers
               of anthropogenic CO_2 emissions. "RCPs are the product of an
               innovative collaboration between integrated assessment modelers,
               climate modelers, terrestrial ecosystem modelers and emission
               inventory experts" (van Vuuren *et al.*, 2011). The four RCPs
               are RCP2.6, RCP4.5, RCP6.0, and RCP8.5, where the number
               appended to RCP represents the global radiative forcing (W/m²)
               in the year 2100. RCP2.6 represents a scenario of heavily curbed
               CO_2 emissions, whereas RCP8.5 represents unmitigated emissions.
               These RCPs are the recommended inputs for the Coupled Model
               Intercomparison Project phase 5 (CMIP5).
Science   The global atmospheric CO_2 concentrations associated with
               the four RCPs are heavily used by the global climate modeling
               community, as the prescribed/input values of CO_2 in the future

DIVE INTO THE DATA BOX 1.3 (cont.)

(for which we have no observations, of course). In future climate change projections, we distinguish scientific uncertainty from societal uncertainty. The different RCPs represent the range of societal uncertainty. In other words, even if we had perfect climate models (or a single climate model), there would still be large uncertainty in many climate change projections because of how different the four RCPs are.

Figure	1.6
Version	2.0.5
Variable	Atmospheric concentration of carbon dioxide
Platform	Scenario development
Spatial	1D (time series), global
Temporal	2005–2100, annual averages
Source	www.iiasa.ac.at/web-apps/tnt/RcpDb
Format	ASCII (.dat) and Excel (.xls)
Resource	co2.m; co2.mat (sample data and code provided on publisher website here: www.cambridge.org/karnauskas)

Journal Reference for Data

van Vuuren, D. P., Edmonds, J., Kainuma, M., *et al.* The representative concentration pathways: an overview. *Clim. Change* **109**, 5–31 (2011). DOI: 10.1007/s10584-011-0148-z.

Sample Journal Article Using Data

Kay, J. E., Deser, C., Phillips, A., *et al.* The Community Earth System Model (CESM) Large Ensemble Project: a community resource for studying climate change in the presence of internal climate variability. *Bull. Amer. Meteor. Soc.* **96**, 1333–1349 (2015). DOI: 10.1175/BAMS-D-13-00255.1.

2 The Heat Budget

2.1 Introduction to the Ocean Mixed Layer Heat Budget

2.1.1 The Nexus of Ocean–Atmosphere Interaction

The surface **mixed layer** of the ocean is the topmost layer in which key properties of seawater, such as temperature and salinity, are approximately constant over depth. The **density** of seawater in the mixed layer, which is a function of temperature and salinity, is less than that of water further beneath the surface (or else the mixed layer would sink!). The thickness of this layer is highly variable across space and time, and the physical processes that control just how deep the mixed layer extends at a given location and how that may evolve over time will be revealed later in this chapter. Suffice it to say that this is the buffer zone through which *all* exchanges of mass and energy between the ocean and atmosphere must pass. The ocean mixed layer can therefore be considered the nexus of ocean–atmosphere interactions, known to play a key role in climate variability. It is for these reasons that the first few chapters of this text focus on the dynamics of this crucial layer of the ocean – the surface ocean mixed layer.

The ocean mixed layer **heat budget** yields a quantitative perspective on what physical processes control the temperature of seawater in the mixed layer. What sets the heat budget apart from the budgets for salt and momentum in the mixed layer? The primary way in which the atmosphere "feels" the ocean is through its strong sensitivity to the **sea surface temperature**, or **SST** for short. While the salinity and velocity of the ocean surface can have a modest impact on the atmosphere, their primary relevance is to the ocean circulation (which is a major component of the climate system, of course). The temperature of the surface ocean, on the other hand, can directly impact wind patterns, overturning cells in the atmosphere, clouds, rainfall, and more. As we will discover, the heat budget encapsulates the potential for such immediate, *two-way* interactions between the ocean and atmosphere. In other words, the heat budget contains explicit avenues not only for atmospheric phenomena (e.g., wind and sunshine) to influence the ocean, but for ocean conditions to influence the atmosphere as well. Moreover, lurking in the depths of the mixed layer is the capacity for the ocean to quietly bank seemingly random variations in the atmosphere (weather), only to withdraw them later as slow but potent doses of climate variation.

2.1.2   Propelling the Global Ocean Circulation

The ocean–atmosphere interactions that are mediated through the ocean mixed layer heat budget span temporal scales from hours to decades, and spatial scales from meters to large fractions of a hemisphere. These heat exchanges, especially as they play out over longer time scales of months to years, ultimately determine the character of important climatic fluctuations that are felt well into the interior of continents.

Setting aside for the moment the profound climatic variations enabled by the ocean–atmosphere **coupling**, why might a physical oceanographer bother worrying about temporal variations in the temperature of seawater of a few degrees centigrade? Density **gradients** – or variations in the density of seawater across latitude, longitude, and depth – are one of the major drivers of ocean circulation. The sinking of cold and salty (therefore relatively dense) surface waters in the North Atlantic Ocean, for example, is part of the propulsion system for the ocean's overturning circulation. If density was constant throughout the world ocean, it would be entirely up to the wind to drive the circulation of the ocean – a fluid body nearly 300 times more massive than the entire atmosphere! Interestingly, ignoring variations in seawater density leads us to a useful approximation that oceanographers do frequently make – more on this in Chapter 6.

Density turns out to be a difficult quantity to measure directly (or *in situ*), but it can be estimated with impressive precision just from knowing the temperature and salinity. Therefore, the circulation of the global ocean is closely tied to the temperature **field** – or the distribution of temperature over latitude, longitude, and depth – which is shaped by the heat budget.

2.1.3   Diagnosing Observations and Predictions

The ocean mixed layer heat budget is just that – a budget. Much like a financial budget, the balance will increase if the sum of the various inputs is greater than the sum of the various outputs. At the bank, we call these deposits and withdrawals. Upon the unsettling discovery of a steadily declining balance, one tends to attempt to **diagnose** such a trend by examining which of the inputs has disappeared or become too small, and/or which of the outputs is new or has become too large. In much the same vein, the heat budget can be employed as a diagnostic framework to understand such questions as:

- Why did SST increase so much in the central equatorial Pacific, but not the eastern equatorial Pacific, during the 2010 El Niño event? (If you were a Galápagos penguin, you'd certainly care!)
- Why don't global climate models predict significant warming in the subpolar North Atlantic Ocean, even by the end of the twenty-first century? (Does this herald an alarmingly abrupt change in the global ocean circulation?)

A third example plays out with real-world data in Section 2.4.2. When you find yourself wondering how global warming will impact the ocean, the heat budget is a very useful place to start. Humans emissions of carbon dioxide go directly into the atmosphere, which changes the atmosphere in many ways, including its ability to intercept longwave radiation and therefore its equilibrium temperature. However, the temperature will not change at exactly the same pace everywhere on Earth; spatial variations in warming can cause uneven changes in the buoyancy of air parcels at the surface and therefore **sea level pressure** gradients, which drive winds that impart a wind *stress* upon the ocean's surface. The impacts of such atmospheric changes on the ocean, and many others, can be examined directly through the ocean mixed layer heat budget. Which physical processes that change the heat content of the surface ocean are increasing enough to explain the observed and model-predicted SST changes? In places where ocean warming is not as great as might be expected, what cooling processes compensate greenhouse warming? In situations where numerical or computer models are being used, these questions can be answered *a posteriori* (after the fact) using the large volume of output data from such models. This is because numerical models use computers to integrate tendency equations very similarly to the heat budget as presented in this chapter to determine whether the temperature should increase, decrease, or stay the same at each modeled moment in time throughout a model simulation. In the real world, determining the "whys" and "hows" requires the availability of diverse observations contemporaneous to temperature itself, but again the ocean mixed layer heat budget provides the diagnostic framework to make those determinations quantitatively and credibly.

2.1.4   Structure of the Ocean Mixed Layer Heat Budget Equation

The heat budget equation will be presented here generally, and each part of it will be elaborated upon throughout the remaining sections of this chapter. The ocean mixed layer heat budget is a partial differential equation (PDE) with an **Eulerian** specification of the fluid's heat content. In other words, it describes how the heat content $H(x,y)$ of a surface layer at a single, specified geographic location (x,y) varies as time t passes (Figure 2.1). It is therefore an ideal framework for understanding why the climatically important variable of SST rises, falls, or remains constant over time.

In general, two types of physical processes can cause H to change over time for a surface layer of the ocean: an exchange of heat with the atmosphere across the air–sea interface, and a movement of heat within the ocean. To first order, we can represent this local balance mathematically as

$$\frac{\partial H}{\partial t} = Q_{net} - \mathbf{V} \cdot \nabla H \qquad (2.1)$$

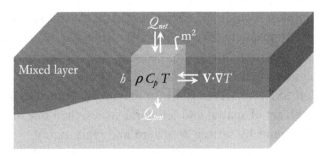

Figure 2.1 Schematic illustration of the ocean mixed layer heat budget. In steady state, and ignoring turbulent mixing, the temperature T or heat content ($\rho\, C_p\, T\, h$) of a volume of water with dimensions 1 m² × h is balanced by surface heat fluxes Q_{net}, solar radiation penetrating through the bottom of the mixed layer Q_{pen}, and temperature advection (both horizontal and vertical) $-\mathbf{V}\cdot\nabla T$.

where Q_{net} is the **net surface heat flux** across the air–sea interface into the ocean mixed layer, $\mathbf{V}$ is the three-dimensional velocity **vector** (u, v, w) and ∇H is the three-dimensional gradient of $H(x,y,z)$.[1] Since heat content H is defined as

$$H = \rho C_p \int_{z=h}^{0} T(z)\,\mathrm{d}z \qquad (2.2)$$

with an assumption of the mean density of seawater ρ (~ 1025 kg m⁻³), the specific heat of seawater C_p (3850 J kg⁻¹ K⁻¹) and knowledge of the depth of the mixed layer h_{ML}, the heat budget equation can be expressed in terms of temperature rather than heat content. For the mixed layer, where temperature is constant throughout the layer by definition, we may replace the integral in Equation 2.2 with simply its product with h_{ML} to yield

$$H_{ML} = \rho C_p T_{ML} h_{ML} \qquad (2.3)$$

and therefore

$$\frac{\partial T_{ML}}{\partial t} = \frac{Q_{net}}{\rho C_p h_{ML}} - \mathbf{V}\cdot\nabla T_{ML} \qquad (2.4)$$

This formulation is especially convenient if one is concerned with the SST as an essential boundary condition to the atmosphere. In fact, we will hereafter drop the subscripts from h_{ML} and T_{ML} such that we imply that mixed layer temperature is equivalent to the SST:

$$\frac{\partial T}{\partial t} = \frac{Q_{net}}{\rho C_p h} - \mathbf{V}\cdot\nabla T \qquad (2.5)$$
$$\quad\ \ \text{A}\qquad\ \ \text{B}\qquad\ \ \text{C}$$

[1] In practice, the treatment of vertical velocity w and the vertical temperature gradient $\partial T/\partial z$ in the ocean mixed layer heat budget diverges slightly from their canonical definitions. This is elaborated upon in Section 2.3.

The heat budget equation (Equation 2.5) implies that if the net surface heat flux between the ocean and atmosphere (B) does not perfectly balance the convergence of heat into the mixed layer from laterally or below by the ocean currents (C), then term A will be nonzero and the SST will change with time.

On the other hand, if SST is observed to be (or defensibly argued to be) constant over some broad enough set of spatial and temporal scales, then one may consider the **steady-state** SST balance by setting A to zero and rearranging:

$$\frac{Q_{net}}{\rho C_p h} = \mathbf{V} \cdot \nabla T \qquad (2.6)$$

$$\text{B} \qquad\qquad \text{C}$$

From this perspective, one can, say, infer how strong the ocean's internal heat transports must be given only the surface fluxes (which are often easier to estimate from observations). Of course, terms B and C of Equations 2.5 and 2.6 are both composed of multiple factors, but the simplicity and broad applicability of the overall structure of the heat budget will not change as we move forward.

2.2   Exchange of Heat with the Atmosphere

As is probably obvious from the previous section, surface heat flux, or the exchange of heat between the ocean and atmosphere directly across their interface, plays a direct role in setting the temperature of some 70 percent of the surface of the Earth and an indirect role for much of the rest of the planet. Heat energy is exchanged across the air–sea interface by fluxes of two types: **radiative** and **turbulent**. The net effect or sum of both of these types of surface fluxes was represented simply by Q_{net} in the previous section (term B in Equations 2.5 and 2.6), and will come into sharper focus here. From here forward, all of the surface fluxes comprising the net surface heat flux Q_{net} will be represented by Q with a subscript denoting the particular heat transfer mechanism involved.

2.2.1   Radiative Fluxes

When a fluid or other substance with mass is not present to serve as a conduit (as in the vacuum of space), energy may still be transmitted in the form of radiation. Two regions of the spectrum of electromagnetic radiation are relevant to surface climate. **Shortwave radiation** emitted by the Sun with wavelength around 0.5 μm (micrometers, or 1×10^{-6} m) reaches the Earth's atmosphere about eight minutes later, entering a veritable maze of reflection, scattering, and absorption. On average, this shortwave radiation reaches and is absorbed by the Earth's surface at a pace of 160–170 W/m². Since 1 watt = 1 joule of energy per second, this is equivalent to 160–170 J being gained by every square meter of the surface every second. Make no mistake, this surface heating by shortwave radiation heating the surface drives the whole show. The rest of the surface heat fluxes, radiative and turbulent alike, are just working to send it back up nearly as fast as it's coming down. This chase for equilibrium (and the inability to reach it completely) sets the ocean and atmosphere into perpetual motion.

The shortwave (or solar) radiation absorbed by the surface will be represented in our ocean mixed layer heat budget equation by Q_{SW}, where the magnitude of this flux into the ocean mixed layer may differ quite considerably from the global average of 160–170 W/m², depending on latitude, season, and cloud cover. A useful functional form is

$$Q_{SW} = Q_0(1-\alpha) - X \tag{2.7}$$

where Q_0 is the solar constant (i.e., the global mean of ~342 W/m² of shortwave radiation making final approach to Earth, after accounting for the sphericity of Earth), α is the **albedo** (the fraction of the incoming shortwave radiation reflected by clouds or the surface, roughly 0.3 for Earth as a whole), and X accounts for absorption of shortwave radiation by stratospheric ozone and other atmospheric constituents. Compared to land, the albedo of the open (ice-free) ocean does not vary much; it is also relatively low (less than 0.1) (Li *et al.*, 2006). So, clouds play a dominant role in determining the value of Q_{SW} reaching the ocean surface at a given place and time, enabling Q_{SW} to change drastically from one day to the next – by as much as 200 W/m² over the tropical oceans! On average, the spatial distribution of Q_{SW} over the global ocean is a strong function of latitude, with the tropics receiving the most; minor variations in this basic structure are primarily a *reflection* (pun intended) of regions with persistent cloud cover (Figure 2.2).

What is the fate of this shortwave radiation entering the ocean? Is all of it actually absorbed by water molecules in the mixed layer and then transformed into kinetic energy, which would raise the water's temperature (by definition)? To accurately account for the potential of Q_{SW} to balance any ongoing cooling processes,

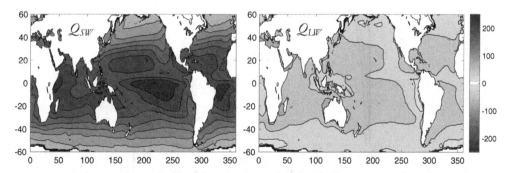

Figure 2.2 Global maps of net surface shortwave Q_{SW} and longwave Q_{LW} radiation (W/m²) averaged from 1984 to 2009 from the WHOI OAFlux data set. Positive (negative) values represent a net flux of radiant energy into (out of) the ocean. Solid lines are contoured every 25 W/m². (A black and white version of this figure will appear in some formats. For the color version, please refer to the plate section.)

we must also account for any shortwave radiation escaping through the base of the mixed layer, or the **penetrative radiation** (Figure 2.1). If it were practical, this quantity would be measured by a waterproof, upward-looking solar radiometer that rises and falls with the base of the mixed layer h. This is not practical, so various methods to estimate the surviving shortwave radiation as a function of depth have been developed that depend on optical characteristics of the water (clarity, biological productivity, suspended inorganic matter, etc.).

In the clearest of seawater, around 75 percent of the entering shortwave radiation is absorbed by 10 m depth, while biologically productive or turbid coastal waters will extinguish around 95 percent by 10 m. Given that the vertical profile of shortwave radiation follows an exponential decay function down the water column of the form

$$Q_{pen} = Q_{SW}e^{-h} \tag{2.8}$$

penetrative radiation Q_{pen} will be larger and more important to apply as an adjustment to Q_{SW} the shallower the mixed layer depth h. Research in the 1970s by oceanographers at the Oregon State University (Paulson and Simpson, 1977) estimated how much h would need to be scaled by in order to properly account for penetrative radiation across a wide range of water types, as defined a couple of decades earlier by Danish oceanographer Nils Jerlov (Figure 2.3). This is our first example of a **parameterization**, which is the process of estimating a quantity that is difficult to measure or compute directly, based on its close relationship to other parameters that are easier to determine.

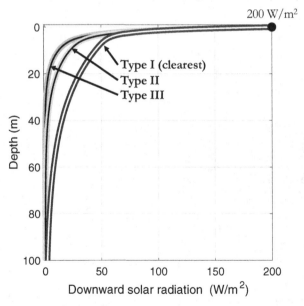

Figure 2.3 Vertical profiles of downward shortwave radiation (W/m²) for three different water types and a net surface shortwave radiation Q_{SW} of 200 W/m². Types I to III represent water types of decreasing clarity, in that order.

The second and arguably simpler of the surface radiative fluxes is **longwave radiation**, which is the emission of infrared or "thermal" radiation by any object with a temperature above absolute zero. Accurate estimation of the longwave radiation Q_{LW} emitted by the surface of the ocean is made possible by the Stefan–Boltzmann Law, which states that the longwave radiation emitted by the sea surface (or any object) is proportional to the absolute SST above absolute zero (i.e., in Kelvin) raised to the fourth power,

$$Q_{LW} = \varepsilon \sigma T^4 \tag{2.9}$$

where ε is the emissivity of the object and σ is the Stefan–Boltzmann constant (5.67×10^{-8} W m^{-2} K^{-4}). The nondimensional parameter ε is not quite a constant and depends fundamentally on the wavelength of radiation in question, as well as some properties of the sea surface, like salinity and SST itself. For our purposes it is sufficiently close to unity (0.94–0.99) (Newman *et al.*, 2005) that we will generally assume it to be so.

Given the relatively simple expression for longwave radiation (Equation 2.9), one can imagine the map of upward longwave radiation at the sea surface looks very much like a map of SST itself, with the large values being *very* large due to the (non-linear) dependence on SST. Of course, the greenhouse effect on Earth (due to both the natural presence of greenhouse gases including water vapor and the anthropogenic contribution of additional carbon dioxide, methane, and nitrous oxide) means that some of the longwave radiation emitted by the ocean will be recycled back into the mixed layer rather than completely escaping Earth's atmosphere for outer space. The downward component of longwave radiation may be based on either direct measurements or estimates given some properties of the overlying atmosphere such as humidity and cloud cover – another example of a parameterization. In the context of the ocean mixed layer heat budget, Q_{LW} is typically taken as the *net* longwave radiation, upward minus downward, such that it represents the net amount of longwave radiation emitted by the ocean surface, thus generally acting to lower the temperature and heat content of the ocean mixed layer.

Longwave radiation is especially interesting in the context of the heat budget, because it depends so strongly on the amount of heat (or T) itself. The nonlinearity of the relationship also means that the incremental increase in longwave radiation ΔQ_{LW} for an incremental increase in SST ΔT will be greater at higher temperatures (Figure 2.4). Since a higher T will result in more loss of heat from the ocean mixed layer by way of longwave radiation, it is a rather potent negative feedback on temperature change, or a so-called "thermostat" mechanism. On average, the spatial distribution of Q_{LW} over the global ocean varies by much less than that of Q_{SW}, but a tendency for the warmer water (e.g., in the tropics but away from relatively cool coastal upwelling regions such as along the west coast of the Americas) to emit more Q_{LW} is evident (Figure 2.2).

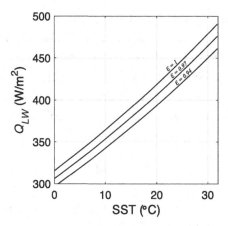

Figure 2.4 Emitted longwave radiation Q_{LW} (W/m²) as a function of SST for three different reasonable values of emissivity ε (1, 0.97 and 0.94).

2.2.2 Turbulent Fluxes

What is notably missing from the makeup of the fluxes introduced in the previous section was any motion whatsoever in either the ocean or the overlying atmosphere. In contrast to radiative fluxes, turbulent heat fluxes are facilitated by fluid motion, in particular motion of the atmosphere relative to the ocean. There are two forms of turbulent heat flux that will be discussed here, and both of them have two essential ingredients: a gradient of a property across the air–sea interface (e.g., a difference of temperature between the ocean surface and that of the air parcel resting on top of it), and a surface wind to keep air parcels moving along before that vertical gradient is diminished through the persistent transfer of that property from the medium with more to the one with less.

Directly measuring turbulent heat fluxes over large regions and over long periods of time, as would be necessary for such measurements to be useful in climate studies, is impractical due to the cumbersome and precise instrumentation necessary, and the spatial scales over which the fluxes themselves can vary. Therefore, **bulk aerodynamic formulae** have been developed, which enable us to quantitatively estimate the values of these fluxes using variables that are easier to observe over large spatial and temporal scales. The bulk formulae for turbulent heat fluxes are quite useful because they relate the magnitude of the fluxes to atmospheric properties including wind speed at a height of 10 m, which is a standard height at which meteorological observations are provided by data products and models. This is another important example of parameterization being a useful and necessary technique in climate studies.

In the case of **sensible heat flux**, the vertical gradient required is that of temperature. Sensible heat flux is equivalent to conduction (of heat) in the classic physical sense that the molecules comprising the two media between which heat is being transferred must be in direct physical contact, which can be treated generally as a diffusive process (Wang and Bras, 1998). However, we will represent it mathematically in the manner consistent with the most common way of calculating it using climate observations and models, i.e., the bulk formula. We express this as

$$Q_{SH} = \rho C_p C_{SH} W \Delta T \tag{2.10}$$

where Q_{SH} is the sensible heat flux in W/m², ρ and C_p are in this case the density and specific heat capacity of the atmosphere, C_{SH} is a transfer coefficient for sensible heat, W is the wind speed at 10 m above the surface, and ΔT is the difference between SST and the surface air temperature. This relation is made possible by the nondimensional bulk transfer coefficient for the transfer of sensible heat flux, C_{SH}, which has been determined empirically from simultaneous atmospheric observations at 10 m to be roughly 1.5×10^{-3}. Good parameterizations are always backed up by rigorous observations, especially to confirm their structure, the values of their coefficients, and the range of conditions under which they apply.

A sensible heat flux from the ocean to the atmosphere will occur where SST is relatively warm compared to the surface air temperature ($\Delta T > 0$), and it will be sustained by the presence of wind ($W > 0$). This turns out to be the case for most of the global ocean, in the annual average. Therefore, sensible heat flux typically acts to cool the mixed layer. Sensible heat flux is most prominent over the major western boundary currents such as the Gulf Stream in the North Atlantic Ocean and the Kuroshio Current in the northwestern rim of the Pacific Ocean (Figure 2.5). Such currents, which will be examined in greater detail in Chapters 4 and 6, transport large volumes of water from the tropics toward the high northern latitudes where air temperatures are much colder than those tropical waters. This is where the ocean–atmosphere temperature differences are largest, and with no shortage of wind in the midlatitudes, there is a ~50 W/m² flux of sensible heat from the ocean to the atmosphere. This localized exchange of sensible heat is one part of why the extratropics on Earth are as habitable as they are, and likewise why the tropics are not excessively warm. Over the remainder of the global ocean, the sensible

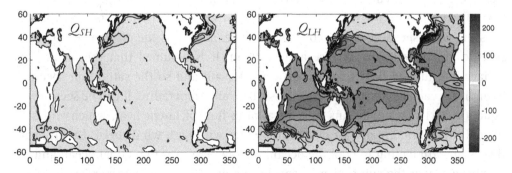

Figure 2.5 Global maps of surface sensible Q_{SH} and latent Q_{LH} turbulent heat flux (W/m²) averaged from 1984 to 2009 from the WHOI OAFlux data set. Positive (negative) values represent a net heat flux into (out of) the ocean. Solid lines are contoured every 25 W/m². (A black and white version of this figure will appear in some formats. For the color version, please refer to the plate section.)

heat fluxes are quite small due to greater thermal equilibrium between the surface ocean and atmosphere, lower wind speeds, or both.

Similar to sensible heat flux, **latent heat flux** depends on a vertical gradient of a property across the ocean and atmosphere, and wind speed. Latent heat flux is the transfer of heat from the ocean to the atmosphere by way of liquid water in the ocean evaporating and entering the atmosphere as water vapor. In order for a kilogram of liquid water in the ocean to change phase and evaporate, it must obtain additional kinetic energy (proportional to the latent heat of vaporization). Where does this energy come from? The heat of the surrounding water molecules in the ocean. It follows, then, that when a kilogram of liquid water evaporates from the ocean surface and enters the atmosphere, the temperature (and volume) of the ocean will decrease and the temperature (and specific humidity q) of the atmosphere will increase. There is a real transfer of mass and energy from the ocean to the atmosphere during evaporation, and the latter is the latent heat flux.

If the atmosphere could hold unlimited moisture, then there would be a very large evaporation and thus latent heat flux everywhere over the ocean at all times, with or without wind. Over time, however, latent heat flux brings the atmosphere closer to saturation (which occurs when relative humidity reaches 100 percent). The closer to saturation, the less latent heat can be transferred because fewer water molecules can evaporate into the atmosphere. In the parlance of mathematics describing diffusion, the reduced gradient of water vapor leads to a lower rate of transfer. Wind speed thus plays a crucial role in turbulent heat fluxes by sweeping away air parcels that are acquiring mass or energy from the ocean and importing "fresh" air parcels that are drier or cooler and therefore able to acquire more water vapor or heat before further eroding the vertical gradient between the air and sea surface. These physical processes are thus encapsulated in the bulk formula for latent heat flux,

$$Q_{LH} = \rho L_v C_{LH} W \Delta q \qquad (2.11)$$

where Q_{LH} is the latent heat flux in W/m^2, ρ is the density of air, L_v is the latent heat of vaporization ($\sim 2.3 \times 10^6$ J/kg), C_{LH} is a transfer coefficient for latent heat, W is the wind speed at 10 m above the surface, and Δq is the difference between specific humidity at the surface and at 10 m. It is assumed that the air at the surface is saturated (it is a *water* surface, after all) and so the saturation specific humidity is calculated using SST as the surface air temperature. It can be seen from Equation 2.11 that conditions ripe for a large flux of latent heat, which will act to cool the ocean, are dry air over warm water with high wind speed. Latent heat flux is upward (acting to cool the ocean) everywhere over the global ocean and is maximized over the subtropical ocean basins due to dry air and persistent trade winds (Figure 2.5). The large latent heat flux over the subtropical oceans leaves an imprint of high sea surface salinity (SSS). Since evaporating seawater leaves the salt behind, salinity is especially concentrated in regions of large latent heat flux.

Comparing the magnitudes of sensible and latent heat fluxes over the global ocean, one can see that sensible heat flux is roughly an order of magnitude smaller than latent heat flux (~10 W/m² compared to ~100 W/m²). The ratio of Q_{SH} to Q_{LH}, known as the Bowen ratio, is used by several disciplines in the Earth sciences to describe the energy flux between phases of water. With a Bowen ratio of ~0.1, the ocean's turbulent energy exchange is said to be dominated by evaporation. The significance of this configuration for atmospheric processes such as circulation and rainfall cannot be overstated. The heating of the lower atmosphere by latent heat flux destabilizes the troposphere (by sharpening its lapse rate). Moreover, the addition of moisture to the atmosphere in places that are then likely to rise and cool means greater potential for additional latent heat release when those water vapor molecules condense, further amplifying the buoyancy of the air parcels and the velocity of the updrafts they are suspended within. This snowball effect fuels a powerful atmospheric process called deep convection, which shapes much of the rainfall and atmospheric circulation, especially in the tropics. Such updrafts tend to pull surface air inward from surrounding regions, shaping the surface winds themselves and leading to convergence zones. Finally, when those convective plumes become sufficiently organized, they might attract the attention of hurricane forecasters as regions likely to develop into tropical cyclones.

2.2.3   Net Surface Heat Flux

To round out our consideration of surface heat fluxes of both the radiative and turbulent variety, we will now examine the global distribution and climatic implications of their sum Q_{net} as

$$Q_{net} = Q_{SW} - Q_{pen} - Q_{LW} - Q_{SH} - Q_{LH} \qquad (2.12)$$

or, substituting with Equations 2.7–2.11,

$$Q_{net} = Q_0(1 - \alpha) - X - Q_{SW}e^{-h} - \varepsilon\sigma T^4 - \rho W(C_p C_{SH}\Delta T + L_v C_{LH}\Delta q) \qquad (2.13)$$

As it turns out, surface fluxes do *not* balance ($Q_{net} \neq 0$) over the majority of the world ocean (Figure 2.6). In other words, longwave radiation and the turbulent heat fluxes are not sufficiently large to balance the incoming solar radiation in some regions, while in other regions they are more than large enough. The subtropics (20–40° latitude in either hemisphere) are dominated by a zonal asymmetry in Q_{net} such that the western boundaries have $Q_{net} < 0$ (losing heat to the atmosphere) and the eastern boundaries have $Q_{net} > 0$ (gaining heat from the atmosphere). The equator, particularly in the Pacific and Atlantic basins, is also dominated by a strong positive Q_{net}. All three of these features are associated with distinct and fundamental components of the ocean circulation. For reasons we will understand

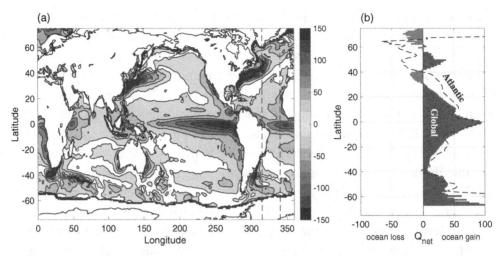

Figure 2.6 (a) Global map of net heat flux into the ocean Q_{net} (W/m²) averaged from 1984 to 2009 using the WHOI OAFlux data set. (b) Profiles of Q_{net} zonally averaged across all longitudes (colored bars) and across only the central Atlantic Ocean (dashed) as marked by dashed lines on the map. (A black and white version of this figure will appear in some formats. For the color version, please refer to the plate section.)

in Chapters 5 and 6, upwelling occurs along the west coasts of continents and along the equator in response to steady winds; why that would cool the surface enough to effectively balance the excess heat is made clear in the following section of this chapter. Also driven in part by steady wind forcing, strong poleward currents known as western boundary currents carry warm water from the tropics to the midlatitudes. Upon reaching the midlatitudes, where the air is much cooler (large ΔT) and drier (large Δq) with winds much faster and more chaotic (large W), sensible and latent heat fluxes together drive a rapid exchange of heat from the ocean to the atmosphere.

On balance, the high-latitude oceans are losing more heat than they are gaining, and the tropical oceans are gaining more heat than they are losing. This is a similar configuration as the global atmosphere and planet as a whole (where the energy exchange is directly with outer space), albeit with far less hemispheric symmetry. In the case of the world ocean, it is the *northern* high latitudes that are losing the most heat (Figure 2.6). Such a latitudinal structure of Q_{net} implies that the ocean must be conveying heat poleward, and this meridional ocean heat transport needn't be hemispherically symmetric; in fact, as we will see in Chapter 7, the Atlantic Ocean exhibits a substantial transport of heat *across* the equator from the Southern Hemisphere to the Northern Hemisphere. It is a fascinating thought exercise to consider what is so different about the Northern and Southern Hemispheres that is needed to explain such hemispheric asymmetries in Q_{net} and the subsequently implied ocean circulation.

2.3 Conveyance of Heat by the Ocean

The previous section examined the myriad ways in which direct heat exchange across the air–sea interface can cause a change in the heat content or temperature of the ocean mixed layer, all of which were encompassed by the term Q_{net}. We now turn our attention to the second half of the right-hand side of the heat budget, or term C in Equations 2.5 and 2.6 in Section 2.1. How do dynamical processes within the ocean itself contribute to changes in its own temperature over time? Equivalently, how can they balance a nonzero sum of surface fluxes ($Q_{net} \neq 0$), such as that shown in the previous section, in order to keep SST at some location close to steady state?

2.3.1 Temperature Advection

Ultimately stemming from the net radiative imbalance on Earth (with the tropics receiving more energy from the Sun than they lose back to space, and vice versa for the poles), the surface ocean is in motion almost everywhere, and there are changes in temperature across space (including depth). We will later return to the specific forces and processes *driving* the ocean's motion, including those arising from interactions with the atmosphere, but here we will take for granted the motion of seawater and simply consider the implications of currents flowing across **gradients** of temperature in the context of the ocean mixed layer heat budget.

Advection is the transport of a fluid's property by the fluid's velocity. This is not to be confused with **diffusion**, which is better thought of as a *spreading* of a fluid's property, always from regions of higher concentration to lower concentration, by random *molecular* motion. Rather, advection relies on the *bulk* motion of the fluid, or the velocity of a *parcel* of water. Virtually any **scalar** property of a fluid can be advected, including temperature T, salinity S, and even the unidirectional components of the vector velocity u, v, and w themselves. The only ingredients necessary for advection are the presence of a gradient of the fluid property and some fluid motion with a nonzero velocity component running parallel to the property gradient (i.e., the projection, or dot product of the velocity vector **V** with the property gradient, does not equal zero).

Consider the case in which the surface heat fluxes perfectly cancel one another, so $Q_{net} = 0$. Term B is eliminated from Equation 2.5, leaving temperature advection as the only remaining process that has been introduced thus far that can cause a change in SST over time at a fixed point in the ocean. Equation 2.5 simplifies to

$$\frac{\partial T}{\partial t} = -\mathbf{V} \cdot \nabla T \tag{2.14}$$

and the dot product can be evaluated on the velocity and gradient vectors to reveal the zonal, meridional, and vertical components of temperature advection:

$$\frac{\partial T}{\partial t} = -u\frac{\partial T}{\partial x} - v\frac{\partial T}{\partial y} - w\frac{\partial T}{\partial z} \tag{2.15}$$

The total temperature advection, and in this case the total change in SST over time, is given by the sum of three terms: the product of the zonal velocity u and the zonal temperature gradient $\partial T/\partial x$; the product of the meridional velocity v and the meridional temperature gradient $\partial T/\partial y$; and the product of the vertical velocity w and the vertical temperature gradient $\partial T/\partial z$. Notice that there is no need to divide these temperature advection components by ρ, C_p, and h as was the case with the Q terms, since the units here are already in kelvin or degrees Celsius per second (m cancels out in the product of m/s and °C/m).

Why there is also a negative sign before each of the three component advection terms, besides simply invoking the definition of the material derivative, is obvious when examining an idealized scenario that includes both of the aforementioned ingredients for temperature advection. Consider a 1000 km² region in the middle of the open ocean in which the surface current is flowing from west to east at 1 m/s and SST is 1 °C warmer along the west side of the square region than the east side. In other words, let $u = 1$ m/s and $\partial T/\partial x = -0.001$ °C/km everywhere; all other velocity components and temperature gradients are zero ($v = w = 0$ and $\partial T/\partial y = \partial T/\partial z = 0$). For any point within this region, we can evaluate the rate of change of SST due to advection as $\partial T/\partial t = -u\,\partial T/\partial x = (-1)\,(1\text{ m/s})\,(-0.000001\text{ °C/m}) = 0.000001$ °C/s = 0.0864 °C/day. Thus, unless there is a negative net heat flux $Q_{net} < 0$ to offset this warming rate due to zonal advection, the SST everywhere within the 1000 km² domain will warm up by about 0.1 °C per day. Mathematics aside, the above example is qualitatively intuitive. Imagine you have tethered yourself to a stationary buoy within the region; you will feel a swift current coming from the west ($u > 0$), and it is warmer to the west ($\partial T/\partial x < 0$). Clearly, you will feel the temperature rise over time under such circumstances, a fact that would not have been represented correctly without the negative sign included in the zonal temperature advection term $-u\,\partial T/\partial x$. Now consider an example with a similar velocity field but a slightly more complex SST field, such that the northwest *corner* of the domain is warmest, and SST decreases in concentric isotherms away from that corner (Figure 2.7). At the location marked A, u is positive (as it is everywhere) but the zonal temperature gradient $\partial T/\partial x = 0$, so there is no horizontal temperature advection at point A. Zonal temperature advection at the location marked B plays out just as in the previous example.

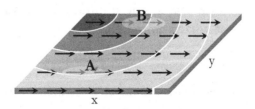

Figure 2.7 An idealized case for evaluating horizontal temperature advection. White contours are isotherms (darker areas warm, lighter areas cold) and black arrows represent the surface velocity field **V** ($u > 0$, $v = 0$).

In principle, there is no distinction between horizontal temperature advection (either zonal or meridional) and vertical temperature advection ($-w\,\partial T/\partial z$), in terms of how they influence the temperature tendency of seawater at some point in three-dimensional space. In practice, however, vertical temperature advection is implemented into the mixed layer heat budget with some nuance, especially in diagnostic studies and numerical models. This is because the mixed layer is by definition isothermal as a function of depth, with a temperature "jump" of some magnitude (toward cooler temperature) immediately beneath the base of the mixed layer h. Depending on the region and season, this jump may be closely related or even synonymous with the **thermocline**, or the transition layer between the relatively warm upper ocean and the colder, abyssal depths. The other practical distinction is that vertical temperature advection generally cannot warm SST – it can only cool. This is because the upper ocean is thermally stratified such that the water from beneath the mixed layer that is transported upward in the case of $w > 0$ is almost always colder than the mixed layer temperature or SST, and in the case of downwelling ($w < 0$), the sinking water is only being replaced laterally due to mass conservation. In one sense, the whole matter is simplified because the math reduces to the temperature difference ΔT between the mixed layer and the water immediately below h multiplied by the vertical velocity exactly at depth h. On the other hand, the mixed layer depth h is not constant, giving rise to additional considerations known as mixing and entrainment, to which we will return.

It is also quite remarkable that horizontal and vertical temperature advection terms often end up with similar orders of magnitude and therefore a similar overall importance to the mixed layer heat budget, but for very different reasons. Consider a typical horizontal temperature gradient $\partial T/\partial x$ or $\partial T/\partial y$ in the tropics, such as the one present in the example above (about 1 °C change in SST over 1000 km). This is 100,000 times more gradual than a typical vertical temperature gradient $\partial T/\partial z$ near the thermocline (about 1 °C temperature change over just 10 m). However, at around 1 m per day, typical upwelling velocities w are also about 100,000 times slower than horizontal velocities u or v, placing vertical temperature advection $-w\,\partial T/\partial z$ on the same playing field as its horizontal counterparts, despite the immeasurably slow (vertical) velocity involved.

Rarely are the velocity and temperature fields of the real ocean as simple as the ones in the previous example. Typically, all three advection terms are nonzero and may even involve different signs (e.g., positive zonal temperature advection and negative vertical temperature advection). The overall time rate of change of SST, or $\partial T/\partial t$, due to advection is the resulting imbalance of the three terms, which by vector identity and as presented earlier in this chapter, is simply the local 3D convergence of heat, or $-\mathbf{V}\cdot\nabla T$.

2.3.2   Diffusion, Mixing, and Entrainment

As alluded to in the previous section, the ocean mixed layer heat budget is not complete without acknowledging that the mixed layer depth h is not constant (through **entrainment** and detrainment), and that the temperature at a specified location can change simply by random molecular motion eroding nearby thermal gradients (the process of diffusion). Further, temperature can change over time due to a process analogous to diffusion, known as turbulent **mixing** by small-scale **eddies**. It is not for lack of importance that these physical processes are relegated to a later section of this chapter. Rather, what they have in common is that they are each something of an enigma. In other words, they can be very challenging to observe *in situ* and to represent mathematically (even parameterize) in universally appealing or meaningful ways. Further complicating matters, both of the specific physical processes of entrainment and diffusion are inseparable from mixing, and some literature equates them under slightly different names. Unsurprisingly, the published research literature in physical oceanography includes a variety of different implementations depending on the data at hand and, under appropriate assumptions, researchers may even choose to neglect mixing. Meanwhile, fundamental theory and observations guarantee us that it does matter in the real ocean and quantitative heat budget analyses will struggle to "close" without accounting for it in some way.

We do not typically account for diffusion at the molecular level; this would indeed be negligibly small and slow compared to the scales of phenomena of interest to physical oceanographers and climate dynamicists. Rather, we use the concept of molecular diffusion as a mathematical analog for representing the effect of turbulent eddy motions to mix water properties such as temperature. The concept derives from Fick's laws of diffusion, which state that (1) the diffusive flux or transport of a solute will be from high to low concentration with a magnitude proportional to the spatial gradient of the solute,

$$F_x = -A\frac{\partial C}{\partial x} \tag{2.16}$$

where F_x is the solute flux in the positive x direction, A is a **diffusivity** coefficient, and C is the solute's concentration; and (2) the time rate of change of the solute's concentration at some location due to diffusive flux will be proportional to the second spatial derivative of the concentration,

$$\frac{\partial C}{\partial t} = A\frac{\partial^2 C}{\partial x^2} \tag{2.17}$$

or in three dimensions:

$$\frac{\partial C}{\partial t} = A\nabla^2 C \tag{2.18}$$

The latter statement (Fick's second law of diffusion) is the basis for our parameterization of ocean mixing, where C becomes a physical property of seawater (rather than the concentration of a solute) such as temperature, and A becomes a measure of the efficiency of turbulent eddies (rather than random molecular motions) to mix such properties. We also underscore the important distinction between true molecular diffusion and the application of its physics to eddies by changing the name of the coefficient A to *eddy* diffusivity. Across the literature, eddy diffusivity is denoted by a variety of letters and symbols, including A, K, k, and κ. We will use K to represent the diffusivity for **tracers** – that is, properties of the seawater other than its velocity, such as temperature and, in the next chapter, salinity. For the local temperature change due to turbulent mixing by eddies, we thus have

$$\frac{\partial T}{\partial t} = K\nabla^2 T \qquad (2.19)$$

which may appear familiar as the classic "heat equation" without a point source. It is also clear from the order of this PDE that temperature change due to mixing ultimately depends not merely on the gradient of temperature, but the second derivative (Laplacian) of the temperature field. This can be rationalized intuitively (albeit overly simplistically) in the following way: We need one derivative to account for the fluid instability likely to give way to turbulence, and another first derivative representing a linear gradient of temperature for the turbulence to erode over time.

If mixing is an important process for tracer budgets in the ocean, which it surely is near the surface, then the million-dollar question is: What is the value of K? The actual value(s) of eddy diffusivity K for the ocean is not well known, and so attempting to deduce it through precise measurements and/or parameterize it well is presently an area of vigorous research. K may well have three different values – one for each spatial dimension (K_x, K_y, and K_z for zonal, meridional, and vertical eddy diffusivity, respectively). K may also have different values for different properties; in fact, when applied to momentum (Chapter 4), this parameter is known as eddy viscosity. Zonal and meridional eddy diffusivities K_x and K_y are most often considered equal to a single horizontal eddy diffusivity K_h, whereas there is good reason to believe that vertical eddy diffusivity K_z can and should be treated differently. When constant values of K_h and K_z are prescribed in published heat budget calculations, they vary substantially, but K_h is always much larger than K_z – by up to eight orders of magnitude (e.g., K_h = 2000 m²/s and K_z = 0.00002 m²/s) (Dong and Kelly, 2004). That $K_h \gg K_z$ reflects the fact that the ocean is stably stratified (by density) in the vertical, and mixing along lines of constant density (isopycnal mixing) requires much less work than mixing across density surfaces (diapycnal mixing).

Rather than a constant, prescribed value for vertical eddy diffusivity K_z being applied to every place and every time, it is commonly preferred to estimate K_z based on the state of the ocean at that particular place and time. A **Richardson number**-conditioned K_z is one example of such a parameterization, where the Richardson number is an estimate of the ratio of the thermal stratification of the fluid to the vertical shear of horizontal flow (and hence the inverse of the propensity for turbulent mixing to occur). Such a parameterized K_z effectively serves the function of one of the vertical derivatives of T in the "Fickian" form of the vertical diffusion, leaving us with the following overall contribution to the ocean mixed layer heat budget by turbulent mixing:

$$\frac{\partial T}{\partial t} = K_h \nabla_h^2 T + K_z \frac{\partial T}{\partial z} \qquad (2.20)$$

where ∇_h^2 is the two-dimensional (horizontal) Laplacian operator $(\partial^2/\partial x^2 + \partial^2/\partial y^2)$. To summarize, even if small-scale turbulent features (such as mesoscale eddies) that lead to strong mixing in the ocean are not resolved in a particular observational data set or numerical model, their overall effect on local temperature tendencies can be parameterized based on aspects of the ocean that typically are observable and/or resolvable by numerical models. In this case, we need only knowledge of the broad-scale temperature field $\nabla^2 T$ and, in the case of Richardson number-dependent vertical diffusivity, the vertical structure of broad horizontal currents $u(z)$ and $v(z)$.

Finally, with recognition that turbulent mixing occurs in the upper ocean and can lead to important variations in SST, we return to the treatment of vertical temperature advection as we left it in the previous section and in Equation 2.15. When mixing occurs at or near the base of the mixed layer h, whether powered internally by turbulent eddy activity as described above or by the influence of strong winds reaching depth h, water from beneath the mixed layer can become entrained into the mixed layer, thereby deepening the mixed layer. Clearly this deepening of h by entraining deeper water will change the temperature of the mixed layer T, since the (enlarged) volume of mixed layer water would then be a combination of the previous (warmer) mixed layer water and the new (cooler) water once residing below the base of the mixed layer.

At depth, vertical temperature advection can be left simply as was defined in Equation 2.15, or $-w\,\partial T/\partial z$. For the special case of the mixed layer, however, we must account for the possible cooling due to entrainment mixing, for which we account with a modified vertical velocity w called entrainment velocity w_e such that

$$w_e = \frac{\partial h}{\partial t} + w_{z=h} \qquad (2.21)$$

where $w_{z=h}$ is vertical velocity at the base of the mixed layer as propelled by, for example, a wind-driven divergence of horizontal volume transports within the mixed layer. The resulting contribution to the mixed layer heat budget therefore only depends on the entrainment velocity w_e and the temperature difference ΔT between the mixed layer and the water immediately below the base of the mixed layer as follows:

$$\frac{\partial T}{\partial t} = -w_e \frac{\Delta T}{h} \qquad (2.22)$$

Two important insights can be gleaned from Equations 2.21 and 2.22. First, since the two terms on the right-hand side of Equation 2.21 are linearly additive, water being entrained from beneath the mixed layer due to a deepening of the mixed layer $\partial h/\partial t$ by 1 m/day is just as effective a cooling mechanism as an actual upwelling velocity of 1 m/day crossing the sharp temperature gradient at the base of the mixed layer. Mixing is important! Second, analogous to the influence of net surface heat flux Q_{net}, the deeper the mixed layer is to begin with, the smaller the impact of entrainment will be, as the cooler water being entrained into the mixed layer will be diluted by a larger initial volume of warm water. On the other hand, for a very shallow mixed layer, the denominator in Equation 2.22 will be small, rendering the temperature response $\partial T/\partial t$ to even a modest upwelling or mixed layer deepening large.

2.4 Putting It Together

2.4.1 Synthesis of Heat Budget Terms and Processes

By elaborating upon the relatively bare-bones equation summarizing the ocean mixed layer heat budget (Equation 2.5), we have accounted for virtually every conceivable reason why the SST, an important climatic variable, may change over time. Combining all of the individual terms in the heat budget as provided above yields what might have otherwise been a daunting equation:

$$\frac{\partial T}{\partial t} = \frac{Q_0(1-\alpha) - X - Q_{SW}e^{-h} - \varepsilon\sigma T^4 - \rho C_p C_{SH}W\Delta T - \rho L_v C_{LH}W\Delta q}{\rho C_p h}$$

$$\text{A} \qquad \text{B} \qquad\quad \text{C} \qquad\;\; \text{D} \qquad\quad \text{E} \qquad\qquad\quad \text{F}$$

$$-u\frac{\partial T}{\partial x} - v\frac{\partial T}{\partial y} - \left(\frac{\partial h}{\partial t} + w_{z=h}\right)\frac{\Delta T}{h} + K_h\left(\frac{\partial^2 T}{\partial x^2} + \frac{\partial^2 T}{\partial y^2}\right) + K_z\frac{\partial T}{\partial z} \qquad (2.23)$$

$$\quad\;\; \text{G} \qquad\;\; \text{H} \qquad\qquad\quad \text{I} \qquad\qquad\qquad\qquad \text{J} \qquad\qquad \text{K}$$

where the top line (terms B–F) represents surface heat exchanges across the air–sea interface and are summarized by Q_{net}, and the bottom line (G–K) represents heat conveyances within the ocean itself. The vocabulary used to define each term is summarized here:

A = SST tendency/time rate of change of the mixed layer temperature
B = solar/shortwave radiation
C = penetrative radiation
D = longwave radiation
E = sensible heat flux
F = latent heat flux
G = zonal temperature advection

H = meridional temperature advection

I = vertical temperature advection and entrainment

J = horizontal eddy diffusion/turbulent mixing

K = vertical eddy diffusion/turbulent mixing.

Accounting for each term in Equation 2.23 is more than just an academic exercise; it is precisely how we can retrospectively understand past observations of ocean and climate variability. It is also how numerical models of the ocean predict tomorrow's SST based on known conditions today. If the sum of all terms on the right-hand side of Equation 2.23 is greater than zero, SST will warm. How might the human influence on the climate system translate into global SST trends (accounting for 71 percent of Earth's surface) of relevance to marine ecosystems and atmospheric circulations, including winds and rainfall? As a reminder, the steady-state heat balance is also borne out when term A is set to zero and terms B–F are set equal to terms G–K. If climate change tweaks term D by trapping more longwave radiation, where will the compensation come from to maintain the equality if not simply from $\partial T/\partial t$ (term A)? Especially at regional scales, there may well be more pathways for climate change to influence the ocean than just the direct greenhouse effect.

In a more general sense, we are now able to see that the ocean mixed layer heat budget is a useful framework for thinking explicitly about the multitude of ways in which the ocean and atmosphere interact, often manifesting as natural climate variability on a variety of time scales. Direct contributions to SST variability by atmospheric processes include cloud cover and stratospheric ozone concentration (term B), downward longwave radiation (implicit in term D), wind speed and air temperature (term E), and humidity (term F). Although not explicitly appearing in terms G–I, the horizontal and vertical velocity components are driven in large part by the wind stress (the dynamics of which we will address in several later chapters), and everywhere that the mixed layer depth h appears, including terms C and I, is a function of the turbulent mixing imparted by the wind stress as well. On the other hand, the SST variations (term A) resulting from any nonzero sum of terms B through K, or the adjustment of any of the terms within B–K playing out in order to keep term A close to zero, can strongly influence the atmosphere. Surface winds and rainfall in the lower troposphere are sensitive to SST anomalies and their spatial gradients, especially in the tropics, which can in turn feed back on almost all of the terms in Equation 2.23. Adjustments of evaporation and latent heat flux (term F) to heat content anomalies arising from other terms can moisten the atmospheric boundary layer and alter its overturning circulations. Variations in the surface ocean's emission of longwave radiation, which will be dramatic with any significant variation in SST given the nonlinear dependence, can have equally large impacts on the atmosphere, especially in warm regimes. The ocean and atmosphere are thus said to be closely **coupled**, because a change in one causes a change in the other and back again. The ocean mixed layer is something of a nexus of the physical processes driving those interactions, and its heat budget helps us account for the exchanges of heat energy. In the next two chapters, we will similarly consider the exchanges and internal processing of freshwater and momentum.

2.4.2 A Real-World Example

The heat budget is a useful framework for conceptually understanding the inner workings of the ocean mixed layer and is therefore a great starting point for appreciating the ocean's role in climate. Applying it quantitatively can be more challenging, in part because of the large number of terms that need to be constrained (i.e., Equation 2.23). Computing each term in the ocean mixed layer heat budget in numerical model simulations is relatively straightforward, since each term is computed explicitly during the model integration and is in fact what the simulated temporal variations in SST are based on. One is effectively guaranteed to find that term A ($\partial T/\partial t$) in Equation 2.23 equals the sum of terms B through K in a model. Doing so with observations requires a range of different variables to be measured simultaneously and with considerable accuracy, which is not always guaranteed to be available at your place or time of interest.

The real-world example we will look at comes from data collected during 2007 at a mooring site in the western equatorial Pacific Ocean (165°E, 0°N). This site is chosen because it is a climatically interesting location – it is situated at the eastern edge of a very large pool of warm water known as the Indo-Pacific (or West Pacific) Warm Pool (Figure 2.8). It is also chosen because a suite of sustained, collocated oceanic and atmospheric measurements have been made at this and an array of sites across the equatorial Pacific Ocean since the 1980s as part of the Tropical Atmosphere–Ocean (TAO) program of the US National Oceanic and Atmospheric Administration (NOAA). Relevant *in-situ* observations collected on the mooring itself include net shortwave and longwave radiation, sensible and latent heat fluxes (computed from bulk formulae using measured wind speed, air temperature, and humidity), zonal and meridional current velocities as a function of depth, SST, and temperature as a function of depth. Horizontal gradients ($\partial T/\partial x$ and $\partial T/\partial y$) cannot be ascertained from a single point, so the values of $\partial T/\partial x$, $\partial T/\partial y$, and $\nabla_h^2 T$ are obtained from satellite observations of SST. The other variable that is not measured directly is upwelling velocity (part of w_e). Upwelling is here taken from a reanalysis known as the Simple Ocean Data Assimilation (SODA), which is a combination of observations and a numerical ocean model.

The satellite observations reveal to us that the 165°E, 0°N TAO mooring is situated along a relatively sharp zonal SST gradient and a broad westward surface current. The mixed layer in this tropical ocean region is quite deep, ranging

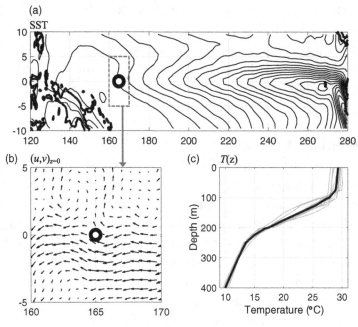

Figure 2.8 (a) Average SST (°C) in the tropical Pacific Ocean during 2007 from the NOAA OIv2 data set. (b) Average surface currents (m/s) in the inset region during 2007 from the OSCAR data set. (c) Vertical profiles of water temperature (°C) at the 165°E, 0°N TAO mooring for each month during 2007 (gray lines) and averaged throughout 2007 (black line). The location of the 165°E, 0°N TAO mooring is marked with a black circle in (a) and (b).

from about 65 m to 125 m based on temperature measurements on the mooring throughout 2007. Here we will define the mixed layer depth as the depth at which the temperature is 0.75 °C cooler than at the surface. This is also an open-ocean region with relatively low biological productivity and hence clear water; the corresponding "Jerlov" water type is IB, which sets the coefficients used in the calculation of penetrative radiation. Finally, let's set the horizontal and vertical eddy diffusivities to 10^3 and 10^{-6} m²/s, respectively.

During the first eight months of 2007, SST at the mooring varied between 29.5 and 30 °C (Figure 2.9). Between August and November, SST cooled by about 1.5 °C, equivalent to a rate of change of SST $\partial T/\partial t$ of –0.5 °C per month. We know from basic thermodynamics that any proposed explanation of such an observation must have, at its core, a net negative energy imbalance. Where did that energy go? To answer that question, we turn to the actual observed (or estimated) values of each term in the ocean mixed layer heat budget and how they stack up against one another. In October 2007, in the midst of the observed cool-down, the latent heat flux, longwave radiation, sensible heat flux, and penetrative radiation observed at the mooring together balanced all but about 100 W/m² of the 284 W/m² of net solar radiation at the surface. Since that represents a *positive* net surface heat flux,

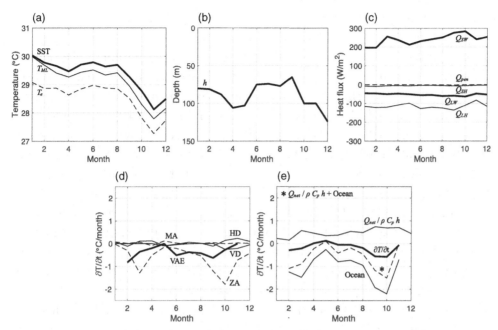

Figure 2.9 (a) SST, temperature averaged throughout the mixed layer (T_{ML}), and entrainment temperature (T_e) during 2007 at the 165°E, 0°N TAO mooring. (b) Mixed layer depth h during 2007. (c) Surface radiative and turbulent heat fluxes during 2007 as observed by the 165°E, 0°N TAO mooring (W/m^2). (d) Ocean terms, or non-surface fluxes, in the ocean mixed layer heat budget including zonal advection (ZA), meridional advection (MA), vertical advection/entrainment (VAE), horizontal eddy diffusion (HD), and vertical eddy diffusion (VD). (e) Summary of heat budget including surface flux terms, ocean terms, their sum, and the actual rate of change of T_{ML} (°C per month).

yet SST was *decreasing*, some processes within the ocean itself must have been operating to decrease the mixed layer heat content at that location.

Among the "ocean" terms in the heat budget, which are always a bit messy compared to the surface flux measurements, meridional advection in October 2007 was negligible, the horizontal and vertical turbulent eddy diffusion terms roughly canceled, and the vertical advection/entrainment term was modest. Without a doubt, negative zonal temperature advection was the physical process driving the overall cooling in October 2007 – contributing a nearly 2 °C per month cooling tendency! Indeed, both the surface current (u) and the zonal temperature gradient ($\partial T/\partial x$) were strongly negative during that month. Given that the mooring is situated precariously along the eastern edge of the Indo-Pacific Warm Pool, one might envisage this as a *contraction* of the warm pool, or equivalently a westward migration of its eastern edge. The overall balance between the surface fluxes (radiative and turbulent) and the ocean processes reveals that the negative zonal temperature advection in October 2007 was more than enough to balance the 100 W/m^2 positive net surface heat flux. These heat budget calculations impute a cooling of nearly −1.5 °C per month, whereas the actual observed cooling rate was only about −0.5 °C per month.

Even with an excellent set of observations, several important uncertainties inevitably remain, including the values for the horizontal and vertical diffusion coefficients (K_h and K_z) and coefficients involved in estimating the transmission of solar radiation through the entirety of the ocean mixed layer (Q_{pen}). Other sources of error lie in the ambiguity of certain definitions, such as how one defines the mixed layer depth h (what is the tolerance on *iso*thermal?) and entrainment temperature T_e being defined as the temperature "immediately" below the base of the mixed layer (how far below h is immediate?). Ocean models, in which the depth of the ocean is divided into a number of discrete "layers" and the topmost layer is typically the mixed layer itself, do not suffer from the latter two issues concerning definitions, whereas models and observations alike are subject to order-of-magnitude level uncertainties concerning mixing. Despite the imperfections in observational data and uncertainties in parameterizations of some physical processes, which often lead to a nontrivial residual (or lack of budget "closure") as was the case in this real-world example, we are able to leverage the physical underpinnings of the ocean mixed layer heat budget to identify the key processes likely responsible for a change at the ocean's surface that surely has implications for the atmosphere and climate – in this case, a 1.5 °C cooling of SST at 165°E, 0°N in the latter half of 2007 symptomatic of a seasonal contraction of the Indo-Pacific Warm Pool.

Questions

1. Briefly define albedo and identify its functional relationship with the ocean mixed layer heat budget.
2. Both the west and east coasts of Australia are regions of substantial net heat loss ($Q_{net} < 0$) for the ocean. Which surface heat flux term contributes the most to the total negative net heat flux in those two regions?
3. Using plain language, why is it necessary to consider a treatment of vertical temperature advection in the ocean mixed layer heat budget that is different from the more general form that might be suitable in the deeper realm of the ocean (i.e., $-w\, \partial T/\partial z$). Write the mixed layer heat budget term in question, and briefly describe each variable.
4. Use Figure 2.3 to estimate the penetrative radiation Q_{pen} in relatively clear water (Type I) with a mixed layer depth h of 20 m, assuming 200 W/m² of shortwave radiation Q_{SW} entering the ocean surface.

5. Imagine a simple planet that does not rotate, with a single, square-shaped ocean basin that has an area of one million square kilometers (i.e., 1000 × 1000 km). The climate of this ocean basin is quite simple – it is sunny all along the western edge, whereas the eastern edge is cloudy. Because of the resultant difference in shortwave radiation Q_{SW} between the western and eastern sides of this ocean basin, the SST is warm in the western half and cold in the eastern half. To be exact, SST along the western edge is 30 °C and SST along the eastern edge is 25 °C. A persistent wind of 7 m/s out of the southwest sets the surface current in motion, reaching a steady-state velocity at the same speed and direction as the wind. What is the rate of change of SST ($\partial T/\partial t$), expressed in units of °C/day, at a point in the center of the basin strictly due to horizontal temperature advection?

6. Physical oceanographers love to debate the best ways to calculate the mixed layer depth in real data. Knowing the subject of the next chapter, can you imagine scenarios where the definition of the mixed layer depth is critically important – why one type of definition or criterion might be better than another? Can you imagine a scenario in which there is, in reality, *no* mixed layer, or an extremely deep mixed layer (hundreds of meters), and what types of physical oceanographic or climatic processes might be occurring in such locations or cases?

7. Select one of the *Dive into the Data* boxes featured in this chapter. Read the sample journal article that uses the associated data set, and describe the general role of that data set in the study. Provide a little context, both technical and scientific: What did the authors actually "do" with the data set? What specific scientific insight was enabled by incorporating this data set into the study?

8. Utilizing *Dive into the Data* Boxes 2.2 *and* 2.3, describe how the equatorial thermocline depth varies as a function of longitude. How does this relate to the variation of SST along the equator?

9. Utilizing *Dive into the Data* Boxes 2.1 and 3.5, can you identify a region of the world ocean that is characterized by roughly this combination of *h* and Q_{SW} in the annual average? Do you think this is also a region characterized by very clear water? (You can utilize *Dive into the Data* Box 6.2 to educate your guess.) Hypothesize on why the region you identified with relatively high Q_{SW} and shallow *h* might be characterized by high or low clarity.

10. Utilizing *Dive into the Data* Box 2.1, calculate and produce a global map of the time-averaged net heat flux Q_{net}. For two different locations in the world ocean, estimate the rate of change of SST ($\partial T/\partial t$) in °C per week that would occur (perpetually) if net surface heat flux constituted the entire ocean mixed layer heat budget (i.e., no advection or mixing). For locations within the tropics, assume a mixed layer depth *h* of 30 m; otherwise, use *h* = 80 m. Assume a constant mixed layer seawater density of 1025 kg/m³.

DIVE INTO THE DATA BOX 2.1

Name	WHOI Ocean–Atmosphere Flux (OAFlux)
Synopsis	Surface heat fluxes (shortwave, longwave, sensible, and latent – as well as their sum, net heat flux) over the global ocean. Satellite observations of relevant quantities are used in standard equations and parameterizations to compute the surface fluxes. The shortwave and longwave fluxes are actually a product of the ISCCP project, but are conveniently distributed alongside the WHOI OAFlux data set. The data set also includes evaporation rate (as a direct byproduct of latent heat flux).
Science	Surface flux data sets like WHOI OAFlux can be used to track variability and trends in air–sea exchanges of heat, applied as boundary conditions in atmosphere and ocean circulation models, and used to diagnose climatic anomalies.
Figures	2.2, 2.5, 2.6, 3.5, 3.6
Version	3
Variable	Surface energy flux
Platform	Satellite
Spatial	Global (gridded), 1° resolution
Temporal	1958–2018, monthly averages*
Source	http://oaflux.whoi.edu/data.html
Format	NetCDF
Resource	oaflux.m; oaflux.mat (sample data and code provided on publisher website here: www.cambridge.org/karnauskas)

Journal References for Data

Yu, L., Jin, X., and Weller, R. A. Multidecade global flux datasets from the objectively analyzed air–sea fluxes (OAFlux) project: latent and sensible heat fluxes, ocean evaporation, and related surface meteorological variables. Woods Hole Oceanographic Institution, OAFlux Project Technical Report. OA-2008-01. Woods Hole, Massachusetts (2008).

Rossow, W. B. and Schiffer, R. A. ISCCP Cloud data products. *Bull. Amer. Meteor. Soc.* **72**, 2–20 (1991). DOI: 10.1175/1520-0477(1991)072<0002:icdp>2.0.CO;2.

Sample Journal Article Using Data

Small, R. J., Bryan, F., Bishop, S., and Tomas, R. Air–sea turbulent heat fluxes in climate models and observational analyses: what drives their variability? *J. Climate* **32**, 2397–2421 (2019). DOI: 10.1175/JCLI-D-18-0576.1.

* Daily averages, 1985–2018 also available. ISCCP shortwave and longwave radiation (and thus net heat flux) only available from 1983 to 2009.

DIVE INTO THE DATA BOX 2.2

Name	NOAA Optimal Interpolation (OI) Sea Surface Temperature (SST)
Synopsis	Blend of *in situ* (various platforms including moorings, ships, etc.) and satellite observations of SST.
Science	Gridded data sets of satellite-observed SST can be used in a wide variety of scientific investigations, such as understanding the spatiotemporal variability of SST from seasonal to decadal time scales, and at regional to global scales. Such data sets can also be applied as boundary conditions to atmospheric general circulation models and used to validate global coupled climate models.
Figures	1.8, 2.8, 2.9
Version	2
Variable	Sea surface temperature
Platform	Satellite, ship, mooring
Spatial	Global (gridded), 1° resolution
Temporal	1982 to present, weekly and monthly averages*
Source	www.esrl.noaa.gov/psd/data/gridded/data.noaa.oisst.v2.html
Format	NetCDF
Resource	oiv2.m; oiv2.mat (sample data and code provided on publisher website here: www.cambridge.org/karnauskas)

Journal Reference for Data

Reynolds, R. W., Rayner, N., Smith, T., Stokes, D., and Wang, W. An improved in situ and satellite SST analysis for climate. *J. Climate* **15**, 1609–1625 (2002). DOI: 10.1175/1520-0442(2002)015<1609:AIISAS>2.0.CO;2.

Sample Journal Article Using Data

Zhao, M. Held, I. M. Lin, S.-J., and Vecchi, G. A. Simulations of global hurricane climatology, interannual variability, and response to global warming using a 50-km resolution GCM. *J. Climate* **22**, 6653–6678 (2009). DOI: 10.1175/2009JCLI3049.1

* Daily averages, 0.25° resolution version also available.

DIVE INTO THE DATA BOX 2.3

Name	NOAA Tropical Atmosphere–Ocean (TAO) Array
Synopsis	An array of moorings was deployed in the tropical Pacific Ocean in the early 1980s, in recognition of the need to monitor and predict climatic variations involving both the tropical ocean and

DIVE INTO THE DATA BOX 2.3 (cont.)

atmosphere, such as El Niño. The TAO Array includes about 50 moorings that are equipped with a variety of atmospheric and oceanic instruments, such as anemometers (for wind), air temperature and humidity sensors, subsurface thermometers and conductivity sensors (for ocean temperature and salinity), and multiple types of ocean current meters (from simple current meters to acoustic ranging instruments). Depending on variable and mooring location, the data records can contain many gaps. The TAO Array is an international effort led jointly by the US NOAA Pacific Marine Environmental Lab (PMEL) and the Japan Agency for Marine–Earth Science and Technology (JAMSTEC).

Science   A great deal of what we know about tropical climate variability and equatorial ocean circulation, not to mention our ability to monitor and predict ENSO conditions, we owe to the TAO Array. The TAO array provides the unique ability to slice through the tropical Pacific Ocean and investigate, both historically and in real time, the interactions between the upper ocean and lower atmosphere.

Figures   1.7, 2.8, 2.9

Version   N/A

Variable   Multiple (e.g., subsurface temperature)

Platform   Mooring

Spatial   Array of moorings spanning the tropical Pacific Ocean

Temporal   Early 1980s to present (with many gaps), variety of averaging from hours to seasons

Source   www.pmel.noaa.gov/tao/drupal/disdel

Format   ASCII (.ascii) and NetCDF (.cdf)

Resource   tao.m; tao.mat (sample data and code provided on publisher website here: www.cambridge.org/karnauskas)

Journal Reference for Data

McPhaden, M. J., Busalacchi, A., Cheney, R., *et al.* The Tropical Ocean–Global Atmosphere (TOGA) observing system: a decade of progress. *J. Geophys. Res.* **103**, 14, 169–14,240 (1998). DOI: 10.1029/97JC02906.

Sample Journal Article Using Data

Moum, J. N., Perlin, A., Nash, J. D., and McPhaden, M. J. Seasonal sea surface cooling in the equatorial Pacific cold tongue controlled by ocean mixing. *Nature* **500**, 64–67 (2013). DOI: 10.1038/nature12363.

3 The Salt Budget

3.1 Introduction to the Ocean Mixed Layer Salinity Budget

3.1.1 A Revolution in Observing our Ocean

In the 1960s, the field of geophysics underwent a revolution with the discovery and rapid understanding of the forces that shape Earth's surface – plate tectonics. At the time of writing this book's first edition, the field of physical oceanography is in the midst of something of a revolution of its own. The first known hydrographic profile in the open ocean was taken in 1751 by Captain Henry Ellis of *The Earl of Hallifax* for what can only be attributed to pure curiosity. Today, there are roughly 4000 autonomous profiling floats covering virtually every swath of the global ocean at a given time, a number that was close to zero only about a dozen years ago. A major objective of the international **Argo** program is to document and understand the heat content and **thermohaline** ("thermo" for temperature and "haline" for salinity) structure of the global ocean, including how it varies over time and space (save for the parts covered by ice – for now). Every 10 days, an Argo float dives to a depth of 2 km and takes a vertical profile of temperature and salinity on its way up (Figure 3.1). Regions of the ocean that once had somewhere between zero and five profiles of temperature and salinity *ever* taken by traditional means (ship-based hydrographic surveys), which was in fact most of the world ocean, now have around 50 profiles and growing.

One of the reasons this is arguably a revolution in physical oceanography is that observing properties beneath the ocean surface is a substantial technical challenge. It is often said that we know more about the surface of the moon than the interior of the ocean; this is probably because the former is easier to access than the latter. Contrast the ocean and the atmosphere. Of course, it is easiest to make observations at the bottom of the atmosphere because that is where gravity has decided we are going to live. But if we want to observe the atmosphere above us, we can probe it with balloons, radars, lidars, and even look at it from above with Earth-orbiting satellites. All of those "remote" sensing techniques enable us to peer *inside* of the atmosphere. The ocean's surface, on the other hand, is remarkably opaque to remote sensing techniques. Over the past few decades, we have developed the capacity to remotely observe (primarily by satellite) several key properties of the surface of the ocean, including its temperature, height, and chlorophyll concentration (a proxy for biological productivity) to an impressive precision, but these only relate indirectly to subsurface properties and processes.

Figure 3.1 Photo of an Argo float being deployed in 2014 by then-graduate student Tyler Hennon on cruise US GO–SHIP P16S in the Southern Ocean under the direction of Chief Scientist Lynne Talley. Associated with the Southern Ocean Carbon and Climate Observations and Modeling (SOCCOM) project, the float pictured is instrumented for biogeochemical measurements such as dissolved oxygen, nitrate, and pH, in addition to the traditional physical variables. Photo credit: Isa Rosso, Scripps Institution of Oceanography. (A black and white version of this figure will appear in some formats. For the color version, please refer to the plate section.)

While it may be too early to guess what the final legacy of this ongoing revolution in ocean observation will be, the science of ocean salinity has certainly been one of the beneficiaries thus far. Argo began filling in our three-dimensional picture of salinity beneath that opaque surface in the early 2000s, while a satellite mission named **Aquarius** filled out our two-dimensional map of sea surface salinity (SSS) from June 2011 through June 2015 (Figure 3.2). Like Argo, the Aquarius mission was an international collaboration, with the Aquarius instrument itself developed and operated by the US National Aeronautics and Space Administration (NASA) and flown aboard the SAC-D satellite vehicle built by the National Space Activities Commission (CONAE) of Argentina. A similar satellite mission named Soil Moisture and Ocean Salinity (SMOS) was also launched by the European Space Agency (ESA) in November 2009.

From what now totals well over one million Argo profiles globally, our knowledge of the distribution of salt in the ocean is filling in nicely beneath the rich maps provided by Aquarius (Figure 3.3). For example, one can identify a broad

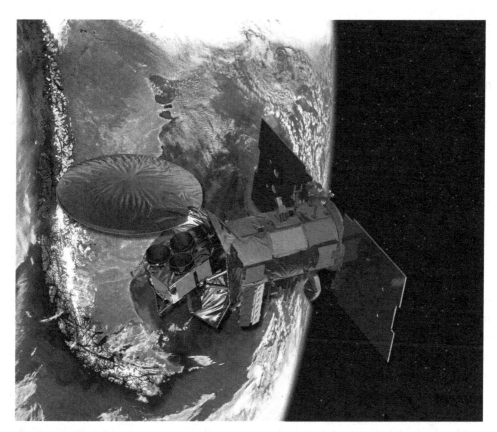

Figure 3.2 Artist's depiction of the Aquarius instrument aboard the SAC-D satellite in orbit. Image credit: NASA. (A black and white version of this figure will appear in some formats. For the color version, please refer to the plate section.)

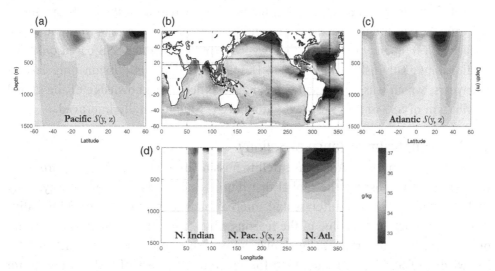

Figure 3.3 Map of SSS (g/kg) from Aquarius satellite observations (b), with cross-sections of subsurface salinity from Argo profiles from 60°S to 60°N through the Pacific (a) and Atlantic (c) Oceans, around the world at ~20°N (d). White areas in the bottom cross-section are landmasses. All data averaged over the same time period (2012–2014). The Roemmich–Gilson Argo Climatology was used. (A black and white version of this figure will appear in some formats. For the color version, please refer to the plate section.)

stream of low-salinity water sinking from the Southern Ocean and spreading northward beneath the salty subtropical gyre – this particular mass of water is called Antarctic Intermediate Water (AAIW) and will be examined in more detail in Chapter 8. Global arrays like Argo have brought to our awareness some profound observations concerning climate change. We knew that a large fraction of the excess heat in the Earth system due to human emissions of greenhouse gases must have been accumulating in the ocean, but it was elusive. Argo gave us the surprising discovery that most of it was hiding in the Southern Ocean – a region with very few observations prior to the Argo program. Ongoing innovations to the Argo platform may eventually enable us to observe the biogeochemistry (carbon, oxygen, nutrients, etc.) of the ocean and even the deep ocean (>2 km) with similar global coverage and resolution.

Unlike the main character of the preceding chapter (SST), our knowledge of the global distribution of SSS was surprisingly blurry before 2011. Prior to Aquarius, the only way to observe SSS was to measure it with an instrument in the water – a precious data point for one time, one place. Aquarius represented SSS catching up to what has been possible for SST since the late 1970s, which is continuous observation with near-global coverage. Why so long? Inferring SST from space is possible by leveraging the Stefan–Boltzmann Law, or rearranging Equation 2.9, which yields a high sensitivity of satellite-measured Q_{LW} to SST variability ($Q_{LW} \propto \sigma T^4$). No such straightforward technique existed for inferring SSS based on the quantities we were capable of remotely sensing with sufficient precision until more recently. Ironically, Aquarius leverages the fact that the emissivity ε of the surface ocean is *not* constant. Emissivity not only depends on SSS, but the magnitude of that dependency varies for different frequencies within the microwave region of the electromagnetic spectrum. Stable and precise measurements of microwave emissions from the ocean surface at a frequency of 1.4 GHz allowed Aquarius to deduce SSS based on those relationships (Koblinsky *et al.*, 2003). Despite the Aquarius mission only lasting four years, which was one year longer than planned, our knowledge of the global distribution of SSS and its seasonal variability is fundamentally improved.

3.1.2   Climatic Importance of Salinity

The half-a-billion dollar Aquarius satellite mission and, in part, the Argo program, was designed to observe the global distribution and variability of salinity. Why such enormous scientific interest and international investment in tracking salinity? Before touching on some of the reasons why salinity is climatically important on its own, it is important to recognize its hidden role in the preceding chapter. Knowledge of SSS is in fact required for accurate diagnosis of SST variability for the following reasons:

- Longwave radiation Q_{LW} is a modest function of emissivity ε, which depends on SSS (Newman *et al.*, 2005). Higher SSS results in lower Q_{LW}.

- Latent heat flux Q_{LH} is a function of the specific humidity gradient across the sea surface Δq, and the saturation vapor pressure of seawater depends on SSS (Sud and Walker, 1997). Higher SSS results in lower Q_{LH}.
- The time rate of change of SST or mixed layer temperature $\partial T/\partial t$ due to a 1 W/m² net surface heat flux Q_{net} depends on the density ρ of seawater, which is a function of salinity. Higher SSS results in a slower SST change.
- The effect of Q_{net} is also damped by the specific heat of seawater C_p, which is a function of salinity. Despite its usual treatment as a constant, C_p varies by approximately 1.4 percent across the typical range of SSS observed in the world ocean (30–40 g/kg) (Sverdrup *et al.*, 1942). In this context, higher SSS results in faster SST change.

Moreover, stratification, as discussed in the preceding chapter in the context of mixing, depends fundamentally on *density*, which is a function of both temperature and salinity. Drawing from the real-world example in the final section of the preceding chapter, this distinction is thought to be a persistent and climatically important feature at the eastern edge of the Indo-Pacific Warm Pool. A **barrier layer** is defined as the layer between the mixed layer depth as defined by density and the **isothermal layer depth** (or, the mixed layer depth as defined by temperature). If the depths at which both temperature and salinity change appreciably are exactly the same, then there is no barrier layer. It turns out this is not always the case, and thus a barrier layer forms that serves to isolate the true mixed layer from the thermocline (Figure 3.4), hiding the cold water from the surface even when mixing or entrainment does occur. Barrier layers owe their existence to the

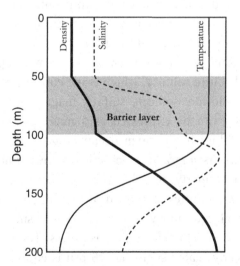

Figure 3.4 Schematic diagram of a barrier layer in hypothetical vertical profiles of temperature, salinity, and density. In this scenario, the isothermal layer depth is 100 m, but the mixed layer depth is only 50 m due to a freshwater lens (low salinity) floating on top of the water column. The barrier layer is therefore found between 50 m and 100 m, with a **barrier layer thickness** of 50 m.

interesting vertical variations in salinity such that a vertical gradient of salinity (and not temperature) is the sole contributor to a vertical gradient in density. In the western equatorial Pacific Ocean, for example, this can be caused by frequent heavy rainfall placing a lens of freshwater upon the surface, where any jump in temperature at the base of that lens is negligible compared to that at the base of the isothermal layer. Many studies have pointed out that both the vertical barrier layer and the sharp horizontal SSS gradient at the eastern edge of the Indo-Pacific Warm (and *fresh*) Pool is crucial for allowing the build-up of heat in the mixed layer prior to major climate anomalies with global reach, such as El Niño events (Maes *et al.*, 2005). As discovered by Janet Sprintall in her dissertation research (Sprintall and Tomczak, 1992), each of the three tropical ocean basins feature barrier layers, and each for very different reasons! More recent studies have also identified patchy synoptic barrier layers in the subtropical oceanic gyres using the well-suited Argo Array (Sato *et al.*, 2006), and others still have uncovered the impact of barrier layers on tropical cyclone intensification (Yan *et al.*, 2017).

Salinity plays an important role in the global ocean circulation. As we will see in later chapters, the ocean circulation can be thought of as two pieces: the part driven by the wind stress and the part driven by density gradients. The latter is known as the thermohaline circulation, which is powered by the sinking of cold, salty waters in the high-latitude regions. Finally, salinity is a valuable indicator of the global water cycle. Section 3.2 will shed light on why we occasionally refer to SSS as the ocean's natural rain gauge.

3.1.3   Structure of the Ocean Mixed Layer Salinity Budget Equation

Before delving into the ocean mixed layer salinity budget, a few definitions are warranted. Unlike temperature, of which the definition and units of measure are generally universal and independent of measuring technique, there are some aspects of defining salinity that are peculiar to oceanography. For an intuitive definition, consider **absolute salinity**, which is the concentration of dissolved salts in seawater such as chloride, sodium, sulfate, magnesium, calcium, and potassium. Absolute salinity can be expressed as a mass fraction (grams of salt per gram of water), and values in the open ocean range from 32 to 37 g/kg. In the late 1970s, oceanographers migrated to **practical salinity units** (or psu) as a more direct function of the property actually being measured in the water, which is electrical conductivity. Salinities expressed on the practical salinity scale are the output of a polynomial equation where the only measured inputs are seawater conductivity ratios, and are therefore dimensionless. The empirically determined coefficients were cleverly scaled such that 35 psu is equivalent to about 35 g/kg. With the adoption of the thermodynamic equation of seawater 2010 (TEOS-10), the international oceanographic community has returned to absolute salinity with the preferred unit of g/kg in an effort to maintain thermodynamic consistency with other variables of interest to physical, chemical, and biological oceanographers. At the end of the day, the numerical differences between absolute and practical salinity units are fairly minor, especially compared to uncertainties in the inference of SSS from space (±0.2 psu). The distinction is meaningful primarily to those who are actually measuring salinity on a boat or in a lab. For interpretation of data, 35 g/kg (or 35 parts per thousand) under the pre-1978 definition of absolute salinity is equal to about 35 psu, which is equal to about 35 g/kg under the TEOS-10 standard. Satellite observations of SSS, such as from Aquarius, tend to use g/kg and psu almost interchangeably. This book will append the units of g/kg to salinity values wherever they are given.

At a basic level, the ocean mixed layer salinity budget is structured similarly to the heat budget covered in the preceding chapter. Rather than heat content $H(x,y)$, it describes how the salt content $S(x,y)$ of a surface layer at a single, specified geographic location (x,y) varies as time t passes. Much like the heat budget, there are fluxes across the air–sea interface and three-dimensional processes within the ocean capable of changing S. Another similarity between the ocean mixed layer heat and salt budgets is that they are both conservation statements; the heat budget applies the conservation of (thermodynamic) energy, whereas the salt budget applies the conservation of mass. This can be seen as follows. The mass of salt M_s contained in a volume of seawater that is 1 m^2 in horizontal size and bounded at the top and bottom by the ocean surface and the mixed layer depth, respectively, is

$$M_s = \rho h S \tag{3.1}$$

where ρ is the mean density of seawater (~ 1025 kg m^{-3}), h is the mixed layer depth (in meters), and S is the absolute salinity (in g/kg). One can see that the units of the product of these three variables works out to grams of salt per square meter (g/m^2). For example, a square kilometer region of the ocean with a depth h of 50 m and salinity S of 35 g/kg holds a mass of salt M_s of about 1.8 million kilograms. Like the ocean mixed layer heat budget, the salt budget states that this quantity need not remain constant at some location, but if it increases it must be coming from somewhere, and if it decreases it must be going somewhere. For some mixed layer depth h, we can express this statement of mass conservation in terms of an Eulerian (local) specification as

$$\underset{A}{\frac{\partial M_S}{\partial t}} = \underset{B}{\rho h \frac{\partial S}{\partial t}} = \underset{C}{F_S + (M_S \text{ import rate} - M_S \text{ export rate})} \tag{3.2}$$

where F_S is the mass flux of salt across the surface and the mass import and export rates are due to the movement of salt within the ocean. An interesting twist in the salt budget is that the salt itself is not actually exchanged between the ocean and atmosphere at a large scale, so F_S is something of a fictitious character until we give it a more physical definition in the next section. The mass import and export rates will take on a familiar form, which we will examine in more detail in Section 3.3.

3.2   Exchange of Freshwater with the Atmosphere

When seawater evaporates, only the freshwater is transferred to the atmosphere; the salt remains in the ocean. Therefore, the surface flux term – term B in Equation 3.2 – is actually driven by the exchange of *freshwater* between the ocean and atmosphere in the form of precipitation and evaporation. This can be thought of as a dilution or concentration of the existing salinity by way of changing the denominator in the definition of absolute salinity, rather than a change in the total mass of salt via the numerator in absolute salinity.

Oceanographers of all disciplines (physical, chemical, biological, geological, and combinations thereof) are far more often interested in the variability of salinity as defined by a concentration of salt, rather than the total mass of salt. The buoyancy of water depends on it, and biology responds to it. So, to express the time rate of change of salinity S, we must define a virtual mass flux of salt as $(E - P) S$ and divide Equation 3.2 by ρh to yield

$$\frac{\partial S}{\partial t} = \frac{(E-P)}{\rho h}S - \mathbf{V} \cdot \nabla S \qquad (3.3)$$

$$ A B C$$

where E is evaporation rate, P is precipitation rate, and term C represents the advection of salinity just as defined for temperature in the preceding chapter (and to be examined further in the next section). A deeper discussion on the physics of precipitation is better saved for a course on atmospheric science, but the factors involved in evaporation are the same as those driving latent heat flux discussed in the preceding chapter. In fact, evaporation rate is equivalent to the latent heat flux simply divided by the latent heat of vaporization of water L_v (~2.3 10^6 J/kg):

$$E = \frac{Q_{LH}}{L_v} \qquad (3.4)$$

E and P are often stated in units like mm/day, which are functionally equivalent to kg m^{-2} s^{-1} (a square meter with 1 mm of water on it has a mass of 1 kg). When their difference is multiplied by the initial salinity S and divided by ρh, the resulting expression quantifies the rate at which S changes over time as the volume of freshwater in the mixed layer varies and either concentrates or dilutes the salt. Therefore, both terms B and C in Equation 3.3 have units of g/kg per second, which is well aligned with our quantity of interest – the time rate of change of salinity $\partial S / \partial t$ (term A).

Throughout the atmospheric, oceanic, and climate sciences, including hydrology and glaciology, the quantity $E - P$ (or $P - E$) is typically referred to simply as **freshwater flux**. Term B in Equation 3.3 as a whole therefore describes the overall effect of ocean–atmosphere exchange of freshwater on salinity. Oceanographers also refer to this as **buoyancy forcing**, since it has a strong local effect on density, and the density of a parcel of water relative to its surroundings determines the buoyancy of that parcel and whether or not it will sink. This buoyancy forcing will reappear in the momentum budget presented in the next chapter, and the global buoyancy-driven flow is the subject of Chapter 8. In case one forgets the structure of term B, it makes intuitive sense that the freshwater flux $E - P$ is multiplied by the initial salinity S, for if salinity was zero (such as a pure freshwater body of water), there could be no future change of S because P only adds more freshwater and E cannot concentrate salt if there is none to begin with. It makes equally good sense that the mixed layer depth h is in the denominator since a deeper h means a larger volume with some initial salinity S, which will be slower to change in response to a given freshwater flux $E - P$ than for a smaller volume with the same initial S.

On average, about 1 m of water is evaporated from the ocean surface per year, and 1 m of water is precipitated onto the ocean surface per year, but the net freshwater flux as defined by their difference $E - P$ is substantial almost everywhere (Figure 3.5). In other words, rainy regions tend to coincide with low evaporation ($E - P < 0$), and regions with high rates of evaporation tend to have

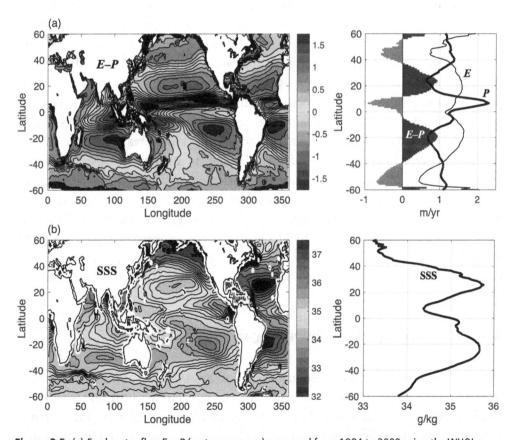

Figure 3.5 (a) Freshwater flux $E - P$ (meters per year) averaged from 1984 to 2009 using the WHOI OAFlux and GPCP data sets and profiles of E, P, and $E - P$ zonally averaged across all longitudes. (b) As in (a), but for SSS (g/kg) averaged from 2012 to 2014 using the Aquarius data set. (A black and white version of this figure will appear in some formats. For the color version, please refer to the plate section.)

low precipitation ($E - P > 0$). In fact, the sign of $E - P$ as a function of latitude is a remarkably robust criterion by which to specify hydrologic regimes, with the **Intertropical Convergence Zone (ITCZ)** (0–10°N) and midlatitude storm tracks (±40–60°) exhibiting a net atmosphere-to-ocean freshwater flux, and the subtropical dry zones (±10–40°) exhibiting a net ocean-to-atmosphere flux of freshwater. The overwhelming dominance of freshwater flux in setting the ocean's climatological salinity is evident; low SSS (~34 g/kg) is found beneath the ITCZ and midlatitudes, while high SSS (~35.5 g/kg) is found in the subtropics. If one looks carefully, however, the zonal mean SSS maxima found in the subtropics of both hemispheres are shifted slightly poleward of the $E - P$ maxima by 2–4° latitude. This curious deviation from the aforementioned dominance of $E - P$ in setting the global SSS field will be explained in the following section.

Not only is SSS a powerful natural "rain gauge" for the ocean, it implies an atmospheric circulation that must be transporting freshwater from regions where the ocean is losing it (the arid subtropics) to regions where the ocean is gaining it – or onto continents where it falls as rain or snow and eventually re-enters the ocean as **runoff**. So robust is the connection between high SSS and freshwater being transferred to the atmosphere as precipitable water vapor ($E - P > 0$) that a handful of recent studies led by Laifang Li and her colleagues exploit observed SSS variability to significantly improve seasonal predictions of rainfall on adjacent continents. The vast and growing archive of salinity profiles from the global Argo Array and other historical observations has also been utilized to document and understand the influence of anthropogenic climate change on Earth's **water cycle**, especially as it plays out over the ocean. The global water cycle response to increased greenhouse gases is of great interest to society, and the ocean is a great place to start since the fundamental patterns are not obfuscated by unevenly distributed features like mountains. Will the regions that are presently dry become even more arid, and the wet get wetter? Or will the arid subtropics expand poleward? One recent study of global salinity trends revealed that SSS has been increasing since 1950 where $E - P > 0$ and decreasing where $E - P < 0$, suggesting an overall intensification of the global water cycle (Durack and Wijffels, 2010).

3.3   Conveyance of Salt by the Ocean

In the preceding chapter, we noted that the ocean must be transporting heat around in order to achieve steady-state balance with the nonzero net heat fluxes with the atmosphere, which were generally positive (meaning the ocean mixed layer gains heat from the atmosphere) in the deep tropics and negative in the high latitudes, with considerable zonal asymmetry (e.g., $Q_{net} < 0$ along the western subtropical boundaries and $Q_{net} > 0$ along the eastern boundaries, at the same latitude). While the rules governing such transports are not fundamentally different when dealing with salinity, it is worth first taking a moment to compare the global map of Q_{net} (Figure 2.6) with its counterpart in the salinity budget ($E - P$) as presented in the previous section (Figure 3.5). One observation is that there is less zonal asymmetry in $E - P$, particularly in the subtropics. Perhaps it is also visually apparent that the map of $E - P$ is more bimodal, meaning a randomly chosen spot on the ocean surface is more likely to be in either a strong positive or negative $E - P$ regime, and less likely to have a local value of $E - P$ near zero. Indeed, the frequency distribution of $E - P$ for the world ocean is markedly bimodal, whereas Q_{net} has a more Gaussian distribution (Figure 3.6). By definition, this means the transition zones (or horizontal gradients) between positive and negative $E - P$ regimes will be quite abrupt. What will the implications be for salinity advection? Let us find out.

The contribution of term C in Equation 3.5 to the time rate of change of SSS, describing the advection of salinity, expands just as it did for temperature into zonal, meridional, and vertical components:

$$\frac{\partial S}{\partial t} = -\mathbf{V}\cdot\nabla S = -u\frac{\partial S}{\partial x} - v\frac{\partial S}{\partial y} - w\frac{\partial S}{\partial z} \tag{3.5}$$

Also as with SST, the vertical advection term $-w\,\partial S/\partial z$ must be treated carefully in the special case of the mixed layer because what really matters is the jump in salinity ΔS that occurs immediately beneath the mixed layer depth h and the entrainment velocity w_e, which includes both the actual vertical velocity at depth h ($w_{z=h}$) and the entrainment (detrainment) of water into (out of) the mixed layer through a change in mixed layer depth $\partial h/\partial t$. The horizontal eddy diffusion term ($K\,\nabla^2 S$) is also included in most diagnostic studies of SSS variability, where 500 m^2/s is a reasonable and often-used value for K, although this is not a well-known quantity and therefore represents a potentially large source of uncertainty.

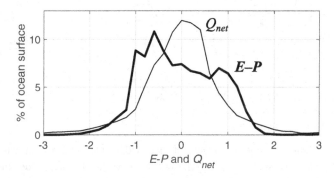

Figure 3.6 Frequency distribution of all values of time-mean Q_{net} and $E - P$ for each 1° × 1° grid cell between 60°S and 60°N. Values of Q_{net} and $E - P$ were normalized by subtracting the global average and dividing by the global standard deviation. Q_{net} and $E - P$ averaged from 1984 to 2009 using the WHOI OAFlux and GPCP data sets.

The complete assembly of non-surface flux terms in the ocean mixed layer salinity budget is therefore

$$\frac{\partial S}{\partial t} = -u\frac{\partial S}{\partial x} - v\frac{\partial S}{\partial y} - w_e\frac{\Delta S}{h} + K\nabla^2 S \qquad (3.6)$$

$$\text{A} \qquad\quad \text{B} \qquad\quad \text{C} \qquad\quad \text{D} \qquad \text{E}$$

where term A is the time rate of change of SSS due to ocean processes, B is zonal SSS advection, C is meridional SSS advection, D is vertical entrainment/mixing, and E is turbulent eddy diffusion.

The implications of the abrupt transitions between $E - P$ regimes are clear in the global distribution of horizontal SSS advection, or the sum of terms B and C in Equation 3.6. The map of $-u\, \partial S/\partial x - v\, \partial S/\partial y$ reveals broad bands of positive and negative SSS advection that stratify by latitude and align nicely with the map of $E - P$ (Figure 3.7). The regions of strong horizontal SSS advection (either positive or negative) correspond directly with the regions of sharp horizontal $E - P$ gradients. Averaging the zonal and meridional SSS advection terms across all longitudes (Figure 3.7) reveals that the meridional SSS advection term $-v\, \partial S/\partial y$ is dominant and generally comparable in size to the freshwater flux contribution to the steady-state salinity balance (the full term B in Equation 3.3, which is plotted as a dashed line in Figure 3.7). The fact that the bands of SSS advection are strongest within ~20° latitude of the equator and are weaker poleward thereof is rooted in the meridional gradients of $E - P$ being strongest within the tropics (at the boundary between the very rainy ITCZ and the very dry subtropics), as well

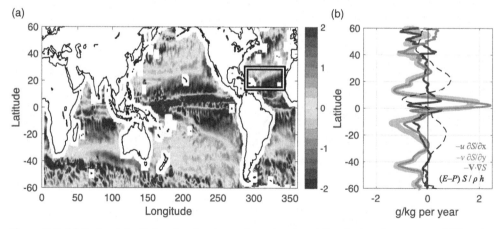

Figure 3.7 (a) Horizontal salinity advection (sum of zonal and meridional advection terms $-u\, \partial S/\partial x$ and $-v\, \partial S/\partial y$) (g/kg per year) using surface velocities from the SODA data set averaged from 1984 to 2008 and SSS from the Aquarius data set averaged from 2012 to 2014. The inset box marks the domain of the Figure 3.8. (b) Profiles of ocean mixed layer salinity budget terms zonally averaged across all longitudes, with spatially varying mixed layer depth in the freshwater flux term based on a fixed density threshold criterion of 0.03 kg/m³. (A black and white version of this figure will appear in some formats. For the color version, please refer to the plate section.)

as meridional velocities v being strongest near the equator – reasons for the latter effect will be discussed in Chapter 6.

We now return to the previously raised question of why the subtropical SSS maxima are observed to be centered a few degrees poleward of the $E - P$ maxima. The answer, which may be unsurprising at this point in the chapter, is that the ocean circulation is pushing it poleward. But how does such a "push" operate through the ocean mixed layer salinity budget? Zooming into the subtropical North Atlantic Ocean, for example, the configuration of SSS gradients and surface currents (one wide SSS maximum in the center with v positive almost everywhere) is such that $-v\, \partial S/\partial y$ is negative south of the SSS maximum and positive north of the SSS maximum (Figure 3.8). Meridional advection therefore tends to decrease SSS immediately south of the maximum and increase SSS immediately to its north, manifested as an overall poleward shift of the SSS maximum relative to where it would have been centered if driven solely by freshwater flux.

Finally, although not exactly a form of ocean conveyance (nor surface flux), it is worth noting three additional processes that contribute to SSS and its time variation in some regions. First, precipitation falling on land and eventually re-entering the ocean as freshwater can substantially lower SSS by dilution. The total amount of **riverine input** to the ocean is estimated to be equivalent to about 10 percent of the rain falling on the entire ocean (Trenberth et al., 2007), which is substantial considering riverine input only occurs at localized points. Near the mouths of major river systems such as the Amazon in the western tropical Atlantic and along the coastlines of the Arabian Sea and Bay of Bengal in the northern Indian Ocean, freshwater plumes are advected horizontally by the surface currents and also mixed and diffused by turbulent eddies. They can be tracked well into the open ocean and have been shown to alter the ocean circulation and even influence monsoon systems by changing surface heat fluxes. Some research applications demand explicitly including this additional source of freshwater flux in the salinity budget along with $E - P$ with a term R (as in river runoff) as $E - P - R$.

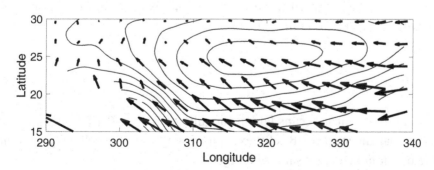

Figure 3.8 Surface currents from the SODA data set averaged from 1984 to 2008 and SSS from the Aquarius data set averaged from 2012 to 2014 in the subtropical North Atlantic Ocean (region outlined by inset box of Figure 3.7).

Second, at high latitudes such as the Arctic Ocean and around the Antarctic continent in the Southern Ocean, **sea ice formation** also factors significantly into the ocean mixed layer salinity budget (Ren *et al.*, 2011). One might think of the process as analogous to evaporation, but for a phase change in the other direction. When sea ice forms, salt is rejected and remains in the ocean. When it melts, fresh liquid water is added to the mixed layer. Where sea ice concentration has large temporal variability, such as seasonal formation and melting, it contributes significantly to the seasonal cycle of SSS and therefore density. Around the fringes of Antarctica, this is known to be an important factor in driving the global ocean circulation as the resulting dense water sinks to great depths; one might theorize how a warming trend would impact such processes.

Finally, **land ice melt** such as glaciers and ice sheets can provide additional point sources of freshwater to the ocean surface in a similar fashion as riverine input. Unlike sea ice, glaciers and ice sheets are not floating on the surface of the ocean and simply ocean water changing phase. Any melting of water from an ice sheet such as the mile-thick one sitting on Greenland represents a new input of freshwater to the ocean that not only raises the sea level but freshens the surface and, once again, can disrupt sinking water that is part of the global thermohaline circulation. Each of these three additional processes initially behaves like a surface flux term, since they are effectively rates of freshwater diluting or concentrating the initial salinity S and are damped by the mixed layer depth h, but can quickly become a major contributor to the "ocean conveyance" terms in the salinity budget by presenting strong gradients to be operated upon by advection, mixing, and diffusion.

Further Reading

Fine examples of seasonal ocean mixed layer salinity budget analyses based on observations include Foltz *et al.* (2004) for the tropical Atlantic Ocean and Dong *et al.* (2009) for the Southern Ocean. For a similar example but based on a global model simulation, see Qu *et al.* (2011).

Questions

1. Are the subtropical oceans (along roughly 20°S and 20°N) relatively fresh or salty near the surface? Briefly explain why, referring specifically to terms in the ocean mixed layer salinity budget.

2. Consider the ocean mixed layer salinity budget. Say salinity is increasing over time at two nearby locations A and B because evaporation is greater than precipitation, and advection is negligible. What characteristic of the upper ocean could be different between the two locations such that salinity is increasing faster at location A than at location B, even with the same $E - P$ difference?

3. Aside from advection, the ocean mixed layer salinity budget is controlled by the difference between evaporation and precipitation. What term in the ocean mixed layer *heat* budget is also directly influenced by evaporation?

4. Examine the structure of both the ocean mixed layer heat and salt budgets, and how evaporation fits into both. Based strictly on these two functional relationships, would you expect the time tendencies of SST and SSS to be positively or negatively correlated? Why?

5. Although the atmosphere doesn't "feel" the ocean's surface salinity, *per se*, provide two examples of why salinity is an interesting quantity in climate science.

6. Select one of the *Dive into the Data* boxes featured in this chapter. Read the sample journal article that uses the associated data set, and describe the general role of that data set in the study. Provide a little context, both technical and scientific. What did the authors actually "do" with the data set? What specific scientific insight was enabled by incorporating this data set into the study?

7. Utilizing *Dive into the Data* Box 3.5, graph and describe the average annual cycle of mixed layer depth (MLD) at two different locations in the world ocean. How do the two MLD criteria (based on temperature or salinity threshold) compare at these sites?

8. Utilizing *Dive into the Data* Boxes 2.2 and 3.5, i.e., only SST and MLD observations, sketch vertical profiles of temperature at two different locations in the world ocean on the same graph, from the surface to 500 m. Identify features of your sketches that are well-constrained by observations. What lacking data or information hindered the certainty in your sketches, especially in terms of how the sketches for the two locations compare to one another?

9. Utilizing *Dive into the Data* Boxes 3.1 and 3.3, compare the change in annual average precipitation from 2013 to 2014 with that of SSS (i.e., using global maps). Identify a few regions where this year-to-year change in precipitation is particularly well reflected in the SSS change. Give a very rough estimate of magnitude of the sensitivity of annual average SSS to annual average precipitation rate ($\Delta SSS/\Delta P$) in units of g/kg per mm/day. Be sure to specify the location you are basing this estimate on.

DIVE INTO THE DATA BOX 3.1

Name	Aquarius Sea Surface Salinity
Synopsis	Estimates of SSS from an instrument that measures the emission of microwave energy from the ocean surface (Aquarius), carried aboard the SAC-D satellite. The mission was relatively short (June 2011 through June 2015 total) but proved that SSS could indeed be measured from space. The Aquarius/SAC-D mission was a joint venture between the USA (NASA) and the Argentinean Space Agency CONAE.
Science	Sea surface salinity is an important variable in oceanography and global climatology. Salinity is one of the two main factors in seawater density, which controls convection and the overturning circulation of the ocean. Sea surface salinity is also a reflection of the global water cycle; rainfall and riverine input decrease SSS by diluting the ocean with freshwater, while evaporation of water from the ocean increases SSS by extracting freshwater from the ocean. While long-term studies of oceanography and the water cycle are hardly feasible with around four years of data, several insights on the global distribution of SSS, impacts of river plumes, and tropical intraseasonal variability were enabled by the data set.
Figures	3.2, 3.3, 3.5, 3.7, 3.8, 8.3
Version	5, CAP, Level 3
Variable	Sea surface salinity
Platform	Satellite
Spatial	Global (gridded), 1° resolution
Temporal	2012–2014, monthly averages
Source	http://podaac.jpl.nasa.gov/dataset/AQUARIUS_L3_SSS_RAINCORRECTED_CAP_MONTHLY_V5
Format	NetCDF (.nc)
Resource	aquarius.m; aquarius.mat (sample data and code provided on publisher website here: www.cambridge.org/karnauskas)

Journal Reference for Data

Koblinsky, C. J., Hildebrand, P., LeVine, D., *et al*. Sea surface salinity from space: science goals and measurement approach. *Radio Sci.* **38**, 8064–8067 (2003). DOI: 10.1029/2001RS002584.

Sample Journal Article Using Data

Gierach, M. M., Vazquez-Cuervo, J., Lee, T., and Tsontos, V. M. Aquarius and SMOS detect effects of an extreme Mississippi River flooding event in the Gulf of Mexico. *Geophys. Res. Lett.* **40**, 5188–5193 (2015). DOI: 10.1002/grl.50995.

DIVE INTO THE DATA BOX 3.2

Name   Roemmich–Gilson Argo Climatology

Synopsis   The Argo platform is revolutionizing oceanography. Argo floats are robotic devices about 1 m long and 14 cm across that are deployed anywhere in the ocean, where they drift with the currents, dive to great depths in the ocean (typically 2 km) while continuously recording measurements ("profiling") of temperature and salinity, and return to the surface every 10 days to beam the data up to a communications satellite and back to ground stations on Earth. There are now nearly 4000 Argo profiling floats in the global ocean. The stream of incoming data is overwhelming; derived products like the Roemmich–Gilson Argo Climatology (a *gridding* of the full database of *profiles*) enable end-users (scientists) quick access to the average, global features of the ocean in a more tractable data set.

Science   Data from an individual Argo float can be extracted and analyzed, but perhaps the most important ways in which the global Argo Array has been used is to better quantify changes in ocean heat content – a key indicator of climate change. The Roemmich–Gilson Argo Climatology is a snapshot of the time-averaged conditions of the ocean in terms of temperature and salinity, which can be used to explore the thermohaline (temperature–salinity) structure of the global ocean and to connect subsurface features with sea surface salinity as measured by, for example, Aquarius. There are now hundreds of peer-reviewed, scientific papers published each year that make heavy use of data from the Argo Array.

Figures   3.1, 3.3, cover

Version   Updated December 2017

Variable   Temperature and salinity

Platform   Float

Spatial   Nearly global (65°S–80°N, surface to 2000 m) (gridded), 1° resolution, vertical spacing 10–20 m

Temporal   Climatology representing 2004 through 2016, monthly averages

Source   http://sio-argo.ucsd.edu/RG_Climatology.html

Format   NetCDF (.nc)

Resource   argo.m; argo.mat; argo_full.mat (sample data and code provided on publisher website here: www.cambridge.org/karnauskas)

Journal Reference for Data

Roemmich, D. and Gilson, J. The 2004–2008 mean and annual cycle of temperature, salinity, and steric height in the global ocean from the Argo Program. *Prog. Oceanogr.* **82**, 81–100 (2009). DOI: 10.1016/j.pocean.2009.03.004.

DIVE INTO THE DATA BOX 3.2 (cont.)

Sample Journal Article Using Data

L'Heureux, M. L., Takahashi, K., Watkins, A. B., *et al.* Observing and predicting the 2015/16 El Niño. *Bull. Amer. Meteor. Soc.* **98**, 1363–1382 (2017). DOI: 10.1175/BAMS-D-16-0009.1.

DIVE INTO THE DATA BOX 3.3

Name	Global Precipitation Climatology Project (GPCP)
Synopsis	Blend of satellite estimates (multiple satellite data sets merged) and *in-situ* (rain gauge) observations of precipitation.
Science	With relatively coarse resolution, GPCP is more useful for global analyses of precipitation patterns than local variability. The benefit is that it is globally complete and leverages the strengths of remote sensing measurements (satellites) and *in-situ* observations (rain gauges) to provide the best possible monthly averages around the world. An abundance of climate science has been enabled by GPCP (and similar data products) such as how modes of tropical climate variability like ENSO, the Indian Ocean Dipole (IOD) and the Madden–Julian Oscillation (MJO) influence rainfall patterns.
Figures	1.8, 3.5, 3.6, 7.6
Version	2.3
Variable	Precipitation rate
Platform	Satellite and rain gauge
Spatial	Global (gridded), 2.5° resolution
Temporal	1979 to present, monthly averages
Source	www.esrl.noaa.gov/psd/data/gridded/data.gpcp.html
Format	NetCDF (.nc)
Resource	gpcp.m; gpcp.mat (sample data and code provided on publisher website here: www.cambridge.org/karnauskas)

Journal Reference for Data

Adler, R. F., Huffman, G., Chang, A., *et al.* The Version-2 Global Precipitation Climatology Project (GPCP) monthly precipitation analysis (1979–present). *J. Hydrometeor.* **4**, 1147–1167 (2003). DOI: 10.1175/1525-7541(2003)004<147:tvgpcp>2.0.CO;2.

DIVE INTO THE DATA BOX 3.3 (cont.)

Sample Journal Article Using Data

Trenberth, K. E., Smith, L., Qian, T., Dai, A., and Fasullo, J. Estimates of the global water budget and its annual cycle using observational and model data. *J. Hydrometeor.* **8**, 758–769 (2007). DOI: 10.1175/JHM600.1.

DIVE INTO THE DATA BOX 3.4

Name	Simple Ocean Data Assimilation (SODA) Reanalysis
Synopsis	A ocean reanalysis is an ocean general circulation model that is anchored by real observations. The method of anchoring, or "constraining," a model is known as data assimilation (i.e., assimilating real data into a model). Oceanic (and atmospheric) reanalyses are important tools/data sets for climate research because we usually do not have observations everywhere and from every time in the past – especially if one is interested in ocean conditions several decades ago or more. Reanalyses essentially fill the gaps, but with a model's physical equations rather than simply interpolating. The SODA reanalysis is one such data product, yielding several commonly used variables related to the ocean, including temperature, salinity, sea surface height, and currents. Ocean reanalyses like SODA should be interpreted as an estimate of past variability rather than true observations.
Science	The SODA reanalysis can be used to examine past variability in physical ocean conditions, how they relate to the atmosphere, and (especially since SODA v2.2.4 extends back to 1871) how the ocean has changed in response to anthropogenic CO_2 emissions. Changes in ocean currents, heat content, and El Niño are typical scientific foci in peer-reviewed papers that make heavy use of the SODA reanalysis. A key difference in the scientific potential between an ocean reanalysis such as SODA and an instrumental SST reconstruction (e.g., NOAA ERSST v.5) is in the ability to investigate dynamics and subsurface features.
Figures	3.7, 3.8
Version	2.2.4
Variable	Multiple (e.g., subsurface temperature)
Platform	Data assimilation
Spatial	Global (gridded), 0.5° resolution

DIVE INTO THE DATA BOX 3.4 (cont.)

Temporal   1871–2008, monthly averages
Source   https://iridl.ldeo.columbia.edu/SOURCES/.CARTON-GIESE/.SODA/.
               v2p2p4/
Format   NetCDF (.nc)
Resource   soda.m; soda.mat (sample data and code provided on publisher
               website here: www.cambridge.org/karnauskas)

Journal Reference for Data

Giese, B. S. and Ray, S. El Niño variability in simple ocean data assimilation
    (SODA), 1871–2008. *J. Geophys. Res.* **116**, C02024 (2011). DOI:
    10.1029/2010JC006695.

Sample Journal Article Using Data

Yang, H., Lohmann, G., Wei, W., *et al.* Intensification and poleward shift of
    subtropical western boundary currents in a warming climate. *J. Geophys.
    Res. Oceans* **121**, 4928–4945 (2016). DOI: 10.1002/2015JC011513.

DIVE INTO THE DATA BOX 3.5

Name   IFREMER Mixed Layer Depth (MLD) Climatology
Synopsis   The depth of the ocean mixed layer can be estimated using vertical
               profiles of temperature, or temperature and salinity (and thus
               density). Scientists at the French institutions Institut Pierre Simon
               Laplace (IPSL) and Institut français de recherche pour l'exploitation
               de la mer (IFREMER), where the data are now hosted, have compiled
               and quality-controlled over five million hydrographic profiles, from
               which they produced a global, gridded climatology of MLD. The
               authors of the study and data set also develop multiple criteria for
               defining the MLD, including one based on a temperature change
               relative to the surface (2 °C) and a density change relative to the
               surface (0.03 kg/m^3). The primary collections of hydrographic
               profiles used in the production of the MLD climatology include the
               World Ocean Circulation Experiment (WOCE) (see Box 8.1) and Argo
               (see Box 3.2). These collections of hydrographic profiles span 1941
               through 2002, but the climatology is more representative of the
               later periods due to the greater hydrographic sampling of the global
               ocean beginning in the 1960s relative to the 1940s and 1950s.

DIVE INTO THE DATA BOX 3.5 (cont.)

Science   The ocean MLD *h* appears in the denominator of many fundamental equations, budgets, and balances in physical oceanography. As such, it is an important mediator of the influence of wind stress, heat flux, mixing, and other physical processes on the properties of the surface ocean. The MLD is also frequently used as a diagnostic of ocean–atmosphere coupling, such that shallow mixed layers may support strong coupling between the surface ocean and the lower troposphere (since the ocean can respond more rapidly to atmospheric forcing, and the atmosphere can, in turn, respond to the changed surface conditions). When information on the MLD is required for observational calculations or interpretations, or model prescriptions, a reference climatology with realistic seasonal and geographic variability, such as the IFREMER MLD Climatology, is very useful.

Figure   3.7
Version   Updated November 2008
Variable   Mixed layer depth (MLD)
Platform   Hydrographic profiles taken from various platforms (ships, profiling floats, etc.)
Spatial   Global (gridded), 2° resolution
Temporal   Climatology representing 1941 through 2002, monthly averages
Source   www.ifremer.fr/cerweb/deboyer/mld/Surface_Mixed_Layer_Depth .php
Format   NetCDF (.nc)
Resource   mld.m; mld.mat (sample data and code provided on publisher website here: www.cambridge.org/karnauskas)

Journal Reference for Data

de Boyer Montégut, C., Madec, G., Fischer, A. S., Lazar, A., and Iudicone, D. Mixed layer depth over the global ocean: an examination of profile data and a profile-based climatology. *J. Geophys. Res.* **109**, C12003 (2004). DOI: 10.1029/2004JC002378.

Sample Journal Article Using Data

Lau, K. M. and Kim, K. M. Cooling of the Atlantic by Saharan dust. *Geophys. Res. Lett.* **34**, L23811 (2007). DOI: 10.1029/2007GL031538.

4 The Momentum Budget

4.1 Introduction to the Ocean's Momentum Budget

4.1.1 Motivation and Linkages to Heat and Salt Budgets

The previous two chapters explored the physics that moderate changes in sea surface temperature (SST) and salinity (SSS) through a consistent budget framework. Both the heat and salt budgets contained essentially two groups of terms that perform a balancing act in steady state: surface fluxes and ocean processes including advection, where the velocity vector **V** is the engine behind the latter. In that context, one may view the present chapter as an appendix to the previous two – a further unpacking of those advection terms. But it is so much more; the momentum budget is the entry point for understanding the global ocean circulation itself. Why, for example, is there a persistently strong northward current off the Atlantic coast of the USA called the Gulf Stream, and a similar feature in the Pacific Ocean? Why is there a strong eastward current about 100 m beneath the surface along the equator in both the Pacific and Atlantic Oceans known as the Equatorial Undercurrent, and why does it decelerate when El Niño events occur? Why does the sinking water in the seas around Antarctica routinely accelerate during austral wintertime? All of these features are manifestations of the two general components of the global ocean circulation. Wind stress and density gradients acting as drivers thereof will be explored in Chapters 6 and 8, respectively. So, while many ocean currents are part of a much grander piece of the climate system than can be detected simply from the local balance of forces, the momentum budget is our means to zoom in on the dynamics of their motion.

Like in the beginning of the previous chapter, it is worth taking a moment to notice the points of contact between the three different budgets being examined in this book (Figure 4.1). In fact, each budget is influenced by the outcomes of both of the other budgets. As you now know, the velocity field **V** plays an important role in advection of heat and salinity, which determines how the T and S fields evolve through time. The velocity field also plays a role in the turbulent diffusion of heat and salt, in the sense that the vertical eddy diffusivity depends on the vertical shear of horizontal currents (by the definition of the Richardson number). Outcomes of the heat budget influence the salinity budget since SST modulates evaporation rate, and outcomes of the salinity budget influence the heat budget since longwave emissivity and latent heat flux depend (albeit slightly) on SSS.

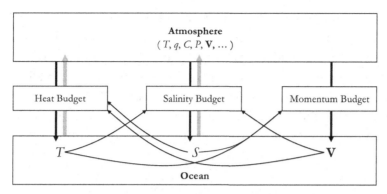

Figure 4.1 A blueprint illustrating connections between the ocean's heat, salt, and momentum budgets (narrow lines) and highlighting the fact that pathways for ocean–atmosphere interaction (heavy lines) are mediated by these three budgets. Example atmospheric variables listed include air temperature (T), specific humidity (q), cloud cover (C), precipitation (P), and the three-dimensional velocity vector ($\mathbf{V}$).

Lastly, as we will notice throughout this chapter, outcomes of the heat and salt budgets influence the momentum budget. The latter point of contact is primarily because temperature and salinity jointly determine density, and density makes a few appearances in the momentum budget. One notable role of temperature in the momentum budget will come, ironically, just as density itself is algebraically eliminated during our derivation of geostrophic currents. A warmer mass of water occupies a greater volume (as it is less dense), a phenomenon that fills the seascape with hills and valleys on which Earth's gravity set water into motion.

Finally, for the student whose primary motivation for learning about physical oceanography is in the context of climate dynamics, it may bring some satisfaction to know that the full spectrum of interactions between the ocean and atmosphere are mediated by these three budgets. The previous two chapters made evident how the state of the atmosphere directly factors into the heat and salt budgets of the ocean; wind stress will ultimately serve that function through the momentum budget. However, even the ocean's influence on the atmosphere takes place through processes encompassed by the heat and salt budgets. Longwave radiation and turbulent heat fluxes cooling the ocean surface also warms the lower atmosphere. Evaporation raising SSS also adds moisture to the lower atmosphere. Thus set in motion are atmospheric perturbations that will eventually influence the ocean on the next go-round. The heat, salt, and momentum budgets are truly the conduits for coupled ocean–atmosphere interaction and therefore global climate variability.

4.1.2 Structure of the Momentum Budget

The ocean mixed layer heat budget (Chapter 2) was a statement of the conservation of energy. The salinity budget (Chapter 3) was an application of the conservation of mass. The subject of this chapter, the momentum budget, is a statement of the conservation of momentum. In fact, it is nothing but a specific application of Newton's second law of motion, $F = ma$. With a simple algebraic rearrangement, we see that Newton's second law of motion states that for a unit mass, acceleration equals the net sum of all forces acting on the body (or water parcel). The greater the mass, the smaller the acceleration for a given sum of forces. That's pretty intuitive! Knowing that the definition of acceleration is a change in velocity **V** over time, we can rewrite Newton's second law of motion for a unit mass as

$$\frac{d\mathbf{V}}{dt} = \sum F \qquad (4.1)$$

It will eventually become necessary to consider forces acting to accelerate seawater in the three Cartesian dimensions separately, so we can rewrite Equation 4.1 as

$$\frac{du}{dt} = \sum F_x \qquad (4.2a)$$

$$\frac{dv}{dt} = \sum F_y \qquad (4.2b)$$

$$\frac{dw}{dt} = \sum F_z \qquad (4.2c)$$

where the subscripts x, y, and z aside the forces F indicate forces acting in the zonal, meridional, and vertical directions, respectively. Recognizing that (1) the three velocity components u, v, and w are scalar quantities; (2) a scalar quantity can be advected by fluid velocity (even velocity itself!) so long as it has a spatial gradient; and (3) advection is not a force, *per se*, we can invoke the definition of the material derivative and add advection to Equation 4.2, yielding

$$\frac{\partial u}{\partial t} = \sum F_x - \mathbf{V} \cdot \nabla u \qquad (4.3a)$$

$$\frac{\partial v}{\partial t} = \sum F_y - \mathbf{V} \cdot \nabla v \qquad (4.3b)$$

$$\frac{\partial w}{\partial t} = \sum F_z - \mathbf{V} \cdot \nabla w \qquad (4.3c)$$

We now find ourselves with a familiar friend: a partial differential equation (PDE) with a local (Eulerian) specification of some property of the fluid – in this case, its velocity. These three equations are a rudimentary form of the classic **Navier–Stokes equations**, widely used to describe the dynamics of viscous flows.

We will also soon find ourselves with a fourth unknown (pressure), which renders the system of equations unsolvable. To enable computers to solve the Navier-Stokes equations on massive scales, we must provide a fourth equation that is deceptively simple but ties the other three together. The **continuity equation** states that the motion of the ocean is nondivergent:

$$\nabla \cdot \mathbf{V} = \frac{\partial u}{\partial x} + \frac{\partial v}{\partial y} + \frac{\partial w}{\partial z} = 0 \qquad (4.4)$$

Implicit in the continuity equation is the assumption that the ocean is incompressible. If there is a convergence of flow in one dimension (say, zonal), then there must be a divergence of flow in one of the other dimensions to balance it. In other words, it is *not* an option that convergence simply "squeezes" more mass into a volume. One can imagine many examples of this principle playing out in the real world. When an eastward wind ($u > 0$) approaches a house, there will be a convergence of zonal momentum ($\partial u/\partial x < 0$) by definition – u must reach zero by the time it hits the building. The continuity equation simply requires that the air must either go around the house (north and/or south, so $\partial v/\partial y > 0$) or up and over it (so $\partial w/\partial z > 0$) in order to balance the zonal convergence. This is clearly an approximation whose validity depends on spatial scale. The propagation of sound in the ocean is *only* made possible by the compressibility of seawater, but we are not concerned about the molecular scale here. While the three component momentum budget equations define the Navier–Stokes equations, the Navier–Stokes equations along with the continuity equation comprise what oceanographers commonly refer to as the **equations of motion**.

Despite the similar functional form of the momentum budget to the heat and salt budgets, there are some key differences to be aware of. The first, of course, is that there are three budget equations (plus the continuity equation) rather than just one. The horizontal momentum budgets (zonal and meridional) will have but minor differences, whereas the meaning of the terms in the vertical momentum budget can be quite different. The second is that the terms in the momentum budget do not separate so cleanly into surface fluxes and ocean conveyance of momentum. There are simply forces, and there need not be any exchange of momentum with the atmosphere for there to be substantial ocean momentum or temporal variability thereof. In the case where there *is* a flux of momentum across the air–sea boundary, it will be in the form of wind stress (an atmosphere-to-ocean flux), whereas there is negligible exchange of momentum from the ocean to the atmosphere – another stark difference from the previous chapters. Finally, the qualifier "mixed layer" does not apply very well to the momentum budget for a couple of reasons. The ocean mixed layer is defined as an isothermal, isohaline, or isopycnal layer at the surface, while velocity may vary considerably throughout the depth of such a mixed layer, beginning with substantial changes in speed and direction within the top meter or so. Horizontal velocities u and v in the advective terms of the heat and salt budgets are therefore best represented by the velocity averaged over the depth of the mixed layer (or some other definition of the top boundary layer), when such data are available. The other, and perhaps fortuitous, reason is that the momentum budget as generally constructed is just as applicable at depth as it is near the surface due to the aforementioned lack of uniquely surface terms (e.g., evaporative heat flux only occurs at the surface, where a phase change is possible). Even wind stress (the frictional force imparted by the atmosphere onto the top layer of the ocean) can be replaced in the horizontal momentum budget seamlessly with the friction imparted by one layer of water onto the one beneath it.

With some context and a general structure now in mind, the next section gives some specific meaning to each force in the momentum budget erstwhile denoted simply as $\Sigma\,F$, while Section 4.3 invokes some justifiable assumptions about the flow that greatly reduce the complexity of the momentum budget and enables us to gain extremely powerful insight into the observed upper ocean circulation.

4.2   Forces and Processes Driving Ocean Currents

Imagine a hypothetical cube of water in the ocean – perhaps one near the surface about 100 km north of the Hawai'ian island of Oahu. Let's also imagine it is a particularly calm day with no wind, and our cube of water is perfectly at rest ($\mathbf{V}$ = 0). Before reading on, ask yourself: What might be capable of setting this cube of water into motion? There are only three fundamental geophysical forces that can act on our cube of water – **gravity**, the **Coriolis force**, and **friction** – and each will appear logically on the right-hand side of Equation 4.3. As we will see, one of these (Coriolis) can only modify existing flow, and friction can come in a multitude of forms, so let us begin with the simplest force – gravity – which is indeed capable of accelerating our resting water parcel in any direction, even if the winds remain calm.

4.2.1   Gravity and Pressure

It is easy to imagine gravity exerting an influence on vertical velocity, but gravity is also a fundamental driver of *horizontal* velocity, especially when masquerading as the **pressure gradient force**. Pressure p is defined as the force F applied per surface area A, so let us consider a container of water with a thin separator inserted exactly down the center of the container (Figure 4.2). Let us also assume that the water in the left half of the tank is filled higher than on the right side of

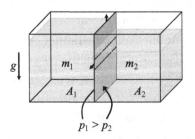

Figure 4.2 Schematic diagram illustrating the role of gravity in driving horizontal fluid accelerations by way of the pressure gradient force.

the separator. Since the surface area at the bottom of the container is the same on either side of the separator ($A_1 = A_2$), the only difference in pressures p_1 and p_2 will be due to the difference in forces F_1 and F_2 being applied. So, how do the forces F_1 and F_2 compare to one another? We can express Newton's second law of motion featuring the acceleration due to gravity g (9.81 m/s²) as $F = mg$, and with g being a constant, the only difference between the forces F_1 and F_2 will be due to the difference in the masses of the water on either side of the separator m_1 and m_2. Since there is a greater volume of water on the left, the mass is greater on the left ($m_1 > m_2$), the downward force (due to gravity) is greater on the left ($F_1 > F_2$), and consequently the pressure exerted is greater on the left ($p_1 > p_2$). Raising the separator by a small increment will result in the water at the bottom flowing from left to right through the gap, and this acceleration of the water that was initially at rest will be proportional to the pressure difference Δp.

Of course, we are not just interested in currents at the very bottom of the sea, so it is important to recognize that the linear dependence of horizontal acceleration on the horizontal gradient in pressure demonstrated through the above thought experiment works just the same at any depth. Imagine that halfway up the container, there is a sliding segment of the separator that can be removed sideways through a slit in the front wall of the container. Halfway up the container, the same volume of water will be *below* that level on either side of the separator, so the *difference* in pressure Δp across the separator at that depth will be exactly the same as it was at the bottom, and hence the horizontal acceleration from left to right through the opened segment will be the same. Driving this concept even further, it may already be obvious that knowing the difference in *height* of the water surface is sufficient to estimate the horizontal acceleration due to the pressure gradient force with impressive accuracy. This is fortunate, since precise global observations of the sea surface height are now routine; we will formally apply such a conversion later in this chapter.

Summarizing generally, then, acceleration is proportional to the local gradient of the pressure field, but of course a denser fluid will be more sluggish to change its velocity in response to the same pressure gradient. We must also introduce a negative sign such that fluid will accelerate from regions of higher pressure toward regions of lower pressure. To express the three component accelerations due to the pressure gradient force in a form that aligns with our momentum budget framework (Equation 4.3),

$$\frac{\partial u}{\partial t} = -\frac{1}{\rho}\frac{\partial p}{\partial x} \tag{4.5a}$$

$$\frac{\partial v}{\partial t} = -\frac{1}{\rho}\frac{\partial p}{\partial y} \tag{4.5b}$$

$$\frac{\partial w}{\partial t} = -\frac{1}{\rho}\frac{\partial p}{\partial z} \tag{4.5c}$$

As can be deduced from Equation 4.5c, there will always be an upward velocity tendency since the overlying volume or mass, downward force, and thus pressure increase almost linearly with depth. The assumption that this perpetual upward pressure gradient force can be exactly balanced by the downward pull of gravity is called the **hydrostatic balance**, which we will recall when deriving geostrophic currents later in this chapter.

Now that we have fleshed out one force in the momentum budget, which is fundamentally due to gravity despite g not surviving the arithmetic, we can move on to another important force that is capable of imposing accelerations in the form of directional changes.

4.2.2   The Coriolis Force

The second force we will consider as part of the ocean's momentum budget is the **Coriolis force**, which ultimately owes its existence to the fact that Earth is roughly a sphere and rotates about an axis. The Coriolis force is unique for a few reasons. In contrast to the pressure gradient force, which is known as a **body force**, the Coriolis force is an **apparent force** because it is only "apparent" relative to a fixed point (or observer) on the rotating Earth. Also unlike the pressure gradient force, the Coriolis force cannot accelerate resting water into motion. In fact, the numerical value of the Coriolis force is zero for resting water and grows proportional to the horizontal speed of the water. Despite these caveats, the Coriolis force is arguably the most fundamental and ubiquitous force involved in the equations of motion applied to Earth's oceans and atmosphere.

The essence of the Coriolis force is best understood from the perspective of the conservation of **angular momentum**, which is a consequence of Newton's laws of motion. The angular momentum for a unit mass is simply the product of its angular velocity ω (about the axis of rotation) and the distance from the axis squared r^2. Let's think about how this applies to Earth. Whether you are standing on the equator or a kilometer from the North Pole, your path in one rotation of Earth about its axis sweeps out all 2π radians (or 360°) per day, so your angular velocity ω is the same in either case. What is obviously different is the so-called moment of inertia term ($\sim r^2$). A person standing on the equator is one full Earth radius from the axis of rotation (6378 km). The angular momentum of the person standing on the equator is therefore considerably larger than one standing near the North Pole.

Let us now consider a fluid parcel in motion and invoke the *conservation* of angular momentum. Imagine our blob of seawater with unit mass in the North Pacific that is traveling toward the North Pole (as propelled by some other force such as inertia or the pressure gradient force). As it travels northward, its distance from Earth's axis of rotation is decreasing (i.e., r^2 is decreasing), but since the conservation law states that the total angular momentum ($r^2\,\omega$) must remain constant, ω must increase. What is it to increase ω but simply to cover more of those 360°

around the Earth in less time! Our blob of seawater will therefore accelerate in the eastward direction – that is, in the same direction as the rotation of the Earth. In other words, it will be making a gradual right-hand turn.

When explaining the Coriolis effect, some teachers will ask you to imagine a cannonball near the equator fired northward and argue that either (a) the cannonball's initially fast eastward velocity (faster than the Earth anywhere north of the canon) or (b) the Earth spinning beneath the cannonball as it travels northward causes the cannonball to appear to curve. While there is some merit to argument (a) by analogy to angular momentum, argument (b) would have the effect of the cannonball appearing to curve in the opposite direction as that of the Coriolis acceleration. Even overlooking that issue, the teacher is secretly hoping you don't ask why a cannonball fired eastward also turns southward, or a cannonball fired westward turns northward – both of which also constitute right-hand turns. Bodies such as cannonballs or blobs of seawater traveling at a constant speed parallel to lines of latitude (i.e., due east or west) are *changing* neither angular velocity nor distance from the axis of Earth's rotation, so we cannot simply invoke the conservation of angular momentum again (in which one variable changes to compensate for a change in the other). Enter centripetal acceleration.

Now imagine that our blob of seawater in the North Pacific is stationary relative to Earth. It is, of course, actually moving eastward along with the rest of the Earth upon which it rests and, like a ball on a string being swung around in a circle, there is an outward **centrifugal force** balanced by an inward **centripetal force** (courtesy of gravity, in the case of the blob of seawater), which maintains its constant distance r from the axis of rotation. The centripetal force F required for a body of mass m moving at a tangential velocity $\mathbf{V}$ to maintain its distance r from the axis of rotation is

$$F = \frac{mV^2}{r} \tag{4.6}$$

which can be rearranged to highlight that the distance r from the axis of rotation must increase if the velocity $\mathbf{V}$ increases, assuming that the centripetal force F can be maintained:

$$r = \frac{mV^2}{F} \tag{4.7}$$

If our blob of seawater in the North Pacific now begins moving eastward relative to Earth (again, either by some other force, or simply continuing with the previous example that ended up turning it eastward), the slightly increased tendency of the blob of seawater to be flung away from Earth (so as to increase r) due to the greater centrifugal force is easily counterbalanced by gravity – ocean currents and even winds are, after all, very slow compared to the speed of Earth's rotation except very near the poles. Therefore, the only reasonable way for r to increase is for the blob of seawater to move to a lower latitude whose circumference is greater – once again, a gradual right-hand turn. In the opposite case – in which the blob of seawater moves toward the west – **V** becomes smaller, so r must decrease, and it requires far less work to simply turn northward toward higher latitudes than to plunge inward against the upward pressure gradient force (or punch through the seafloor, in the case of a deep current).

Despite how diverse the explanations for the Coriolis force may seem, its implementation into the momentum budget is astonishingly simple. We must merely include a term that accounts for the perpetual "rightward" acceleration in the Northern Hemisphere, and "leftward" acceleration in the Southern Hemisphere, and have it be linearly proportional to the velocity itself. To express the two component accelerations due to the Coriolis force in a form that plugs into our momentum budget framework (Equation 4.3),

$$\frac{\partial u}{\partial t} = fv \tag{4.8a}$$

$$\frac{\partial v}{\partial t} = -fu \tag{4.8b}$$

$$f = 2\Omega \sin\theta \tag{4.8c}$$

where f is the Coriolis frequency, Ω is the rotation rate of the Earth, and θ is latitude. So long as θ is positive (i.e., for the Northern Hemisphere), an eastward acceleration will be induced by a northward velocity (from Equation 4.8a) and a southward acceleration will be induced by an eastward velocity (from Equation 4.8b). In other words, the Coriolis force always accelerates moving fluids to the right in the Northern Hemisphere and to the left in the Southern Hemisphere. This is one of the only times in any branch of Earth science where "right" and "left" are preferred to the cardinal directions, but these rules are foolproof. Rather unremarkably, if f was a constant, then Equations 4.8a and 4.8b would represent a coupled system of equations that draws a perfect circle.

As is clear from the definition of the Coriolis frequency (Equation 4.8c), however, Coriolis accelerations will be stronger at higher latitudes than near the equator given the same velocity u or v. This can also be rectified with the explanations of the Coriolis effect in terms of both angular momentum conservation and centripetal acceleration; in both cases, it simply comes down to the geometry of a sphere. The difference in angular momentum (due to differences in r) between the equator and 1°N is very small compared to the difference in angular momentum between 60°N and 61°N, even though both pairs of latitudes are separated by the same horizontal distance (roughly 111 km). So, as a blob of seawater moves northward near the equator, there is hardly any Δr to be compensated for by $\Delta\omega$, while at higher latitudes, r decreases more rapidly per degree of latitude (or kilometer) and so ω must also increase more rapidly (i.e., turn harder) to conserve angular momentum. Similarly, a unit vector pointing due eastward near the equator peels away from the surface much less steeply than one at 60°N. Therefore, r

must change more rapidly (by veering northward or southward) for a given zonal velocity at high latitudes than near the equator. That the magnitude of the Coriolis frequency depends on latitude has some very important consequences ranging from planetary wave propagation to setting the large-scale structure of the wind-driven ocean circulation. When the Coriolis frequency reappears in our discussion of those consequences in Chapter 6, it will do so by the nickname **planetary vorticity**, so as to distinguish it from relative, absolute, and potential vorticity – another important conserved quantity.

4.2.3   Friction and Wind Stress

The remaining of the three fundamental geophysical forces that can act to accelerate (or change the momentum of) a mass of seawater is friction. Friction in the ocean comes in many forms, from the smallest scale of molecular viscosity to the mixing effects of turbulent eddies. Friction also accounts for the direct transfer of momentum from the wind to the surface ocean, and how that energy is propagated downward to further accelerate water at depths of hundreds of meters and more.

It is not difficult to imagine why viscosity matters. Informally, it is how gooey a fluid is. Formally, it is a measure of the internal friction of the fluid – how much shear stress (or force applied parallel to the surface, per unit area) it takes to make two layers of a fluid move relative to one another. True molecular viscosity v is quite small for seawater ($\sim 10^{-6}$ m^2/s) and therefore inefficient for transferring momentum, so it is almost always ignored in studies of the ocean's momentum budget. However, it is once again the process by which we, by analogy, parameterize the net effect of turbulent eddies on the momentum budget.

It is first worth noticing the close parallels between how friction transfers momentum around within the ocean and how turbulent mixing acts to cause tracer properties such as temperature and salinity to change with time. Mathematically, they are very similar constructs, both phrased as a Fickian diffusion (i.e., the time rate of change of X being proportional to the product of an efficiency coefficient and the second spatial derivative or Laplacian of X). In Chapters 2 and 3, what we called eddy diffusivity (denoted by K) to represent the efficiency of turbulent eddies to mix tracer properties, we call **eddy viscosity** for the mixing of momentum. To be clear, diffusion is the mathematical model in both cases, and viscosity is nothing but diffusivity in the specific case of momentum mixing – they are both simply names of efficiency coefficients.

In the momentum budget, it is necessary to represent horizontal and vertical friction separately for a couple of reasons. Similar to eddy diffusivity for tracers, the coefficients of eddy viscosity are probably very different for horizontal and vertical diffusion. Additionally, wind stress is a major driver of ocean circulation and operates uniquely through the vertical friction term. Before introducing wind stress, then, we can represent the two friction terms of the horizontal momentum budget as

$$\frac{\partial u}{\partial t} = A_h \nabla_h^2 u + A_z \frac{\partial^2 u}{\partial z^2} \qquad (4.9a)$$

$$\frac{\partial v}{\partial t} = A_h \nabla_h^2 v + A_z \frac{\partial^2 v}{\partial z^2} \qquad (4.9b)$$

where A_h and A_z are the horizontal and vertical eddy viscosity coefficients, respectively, and ∇_h^2 is the two-dimensional (horizontal) Laplacian operator $(\partial^2/\partial x^2 + \partial^2/\partial y^2)$. When they are prescribed as constants in numerical models or diagnostic studies, the horizontal eddy viscosity coefficient A_h is typically several orders of magnitude larger than its vertical counterpart A_z. From the abundance of second-order partial derivatives in Equation 4.9, one can reasonably guess that it is not how rapidly the velocity field changes *linearly* with depth or horizontal distance that governs how much it is mixed by turbulent eddies. A linear change in meridional velocity v over zonal distance x (i.e., $\partial v/\partial x = C$ where $C \neq 0$), for example, implies no meridional acceleration by horizontal diffusion. Only where the local slope $\partial v/\partial x$ is itself changing as a function of longitude will the horizontal diffusion term induce a meridional acceleration. Finally, friction terms in the vertical momentum budget are neglected; typical vertical velocities in the ocean are already very slow by comparison to horizontal flow, rendering their higher-order spatial derivatives (i.e., $\nabla^2 w$) virtually impossible to measure directly, prohibitively difficult to parameterize or infer observationally with confidence, and probably not even worth the headache to model numerically, especially when it comes to diagnosing the momentum budget at spatial and temporal scales that matter to climate.

While the wind speed matters to the ocean in some ways such as modulating the surface turbulent heat fluxes and evaporation rate (as discussed in Chapters 2 and 3), it is the closely related wind *stress* that imparts momentum from the air to the sea. The **wind stress**, or the shear stress imposed on the ocean by surface winds, enters the ocean through the vertical friction term of the horizontal momentum budget. In practice, wind *stress* is another difficult quantity to measure directly, but is known to be a robust, nonlinear function of wind *speed* such that

$$\tau_x = \rho_a C_d u_a^2 \qquad (4.10a)$$

$$\tau_y = \rho_a C_d v_a^2 \qquad (4.10b)$$

where τ_x and τ_y are the zonal and meridional components of wind stress, ρ_a is the density of air at the surface, C_d is a drag coefficient, and u_a and v_a are the zonal and meridional wind components at some reference height (historically, the height of a ship deck). Fudginess in C_d notwithstanding, it can be seen from Equation 4.10 that this functional form of wind *stress* (as the square of wind speed) has the effect of exaggerating spatiotemporal differences in surface winds (Figure 4.3). Where wind speed is low, wind stress will be very low relative to how high wind stress is wherever wind speed is high, and the impact of temporal changes in wind speed at one location will be amplified in the horizontal momentum budget for that location.

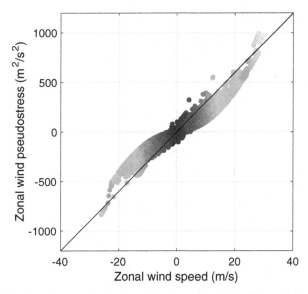

Figure 4.3 Illustration of the relationship between wind speed and wind stress (using pseudostress [m²/s²] from the Cross-Calibrated Multi-Platform [CCMP] level 3.5a data set). Each point on the graph represents one ¼° grid cell over the world ocean for one of the 12 months of the year 2000. There are a total of ~6.5 million points. (A black and white version of this figure will appear in some formats. For the color version, please refer to the plate section.)

If wind stress is known or can be imposed as a boundary condition, then it can be introduced directly into the horizontal momentum budget by replacing the vertical friction terms, as phrased in terms of the detailed vertical structure of u and v in Equation 4.9, such that

$$A_z \frac{\partial^2 u}{\partial z^2} = \frac{1}{\rho} \frac{\partial \tau_x}{\partial z} \tag{4.11a}$$

$$A_z \frac{\partial^2 v}{\partial z^2} = \frac{1}{\rho} \frac{\partial \tau_y}{\partial z} \tag{4.11b}$$

where τ represents the wind stress at the surface or, equivalently, the shear stress at depth. Whence cometh this mysterious equivalence between Fickian-style vertical eddy diffusion and a partial derivative of shear stress? Equation 4.11 in fact derives from the formal yet intuitive definition of viscosity itself (as also articulated in Newton's law of viscosity)! If viscosity takes the form

$$\nu = \frac{\tau_x}{\rho} \frac{\partial u}{\partial z} \tag{4.12}$$

i.e., proportional to the stress required to propel a fluid parcel 1 m/s faster than the fluid parcel 1 m below it, then

$$\frac{\tau_x}{\rho} = \nu \frac{\partial u}{\partial z} \tag{4.13}$$

and, differentiating both sides with respect to z,

$$\frac{1}{\rho}\frac{\partial \tau_x}{\partial z} = \frac{\partial}{\partial z}\left(v\frac{\partial u}{\partial z}\right) \tag{4.14}$$

or assuming v is constant,

$$\frac{1}{\rho}\frac{\partial \tau_x}{\partial z} = v\frac{\partial^2 u}{\partial z^2} \tag{4.15}$$

and thus Equation 4.9 may be rewritten with the divergence of shear stress constituting the vertical friction terms, such that

$$\frac{\partial u}{\partial t} = A_h \nabla_h^2 u + \frac{1}{\rho}\frac{\partial \tau_x}{\partial z} \tag{4.16a}$$

$$\frac{\partial v}{\partial t} = A_h \nabla_h^2 v + \frac{1}{\rho}\frac{\partial \tau_y}{\partial z} \tag{4.16b}$$

As will be seen in Chapter 6, it is even more insightful to consider the depth-integrated horizontal momentum budget where the upper ocean **volume transports** are diagnosed and, in that case, luckily, we need only to evaluate $\tau_x/\rho - A_z \left.\partial u/\partial z\right|_{z=-H}$ and $\tau_y/\rho - A_z \left.\partial v/\partial z\right|_{z=-H}$ at the surface as the vertical friction terms, and $\left.\partial v/\partial z\right|_{z=-H}$ is typically negligible. With such a workable representation of the wind now in hand and incorporated seamlessly into the ocean's momentum budget, the next chapter will make a brief digression into the atmosphere so that we may better understand the setup of the global wind stress field (including its coupling with the ocean – particularly via SST) before returning to apply it to the ocean's momentum budget to understand, aided by some well-justified simplifications, the large-scale wind-driven ocean circulation.

4.2.4   Advection of Momentum

It may at first seem odd that momentum (or velocity) could be advected by velocity, but why shouldn't it? Velocity components u, v, and w are just scalar quantities like temperature or salinity. Particularly in regions of strong horizontal or vertical shear, velocity advection can be an important part of the overall momentum budget. Following Equation 4.3, velocity advection contributes to the momentum budget as

$$\frac{\partial u}{\partial t} = -\mathbf{V}\cdot\nabla u \tag{4.17a}$$

$$\frac{\partial v}{\partial t} = -\mathbf{V}\cdot\nabla v \tag{4.17b}$$

$$\frac{\partial w}{\partial t} = -\mathbf{V}\cdot\nabla w \tag{4.17c}$$

and expanding the advection of zonal velocity (Equation 4.17a), for example, yields

$$\frac{\partial u}{\partial t} = -u\frac{\partial u}{\partial x} - v\frac{\partial u}{\partial y} - w\frac{\partial u}{\partial z} \tag{4.18}$$

From here, we can see that each of the three component momentum budget equations (zonal, meridional, and vertical) will contain nonlinear terms, which would be challenging to solve analytically, but oceanographers working with real data and numerical models are usually approximating spatial gradients by finite difference anyway.

The global distribution of surface momentum advection is relatively patchy (Figure 4.4), but still understandable from first principles. Consistent with its mathematical definition (Equation 4.17), horizontal momentum advection is bound to be significant where there is a confluence of sharp horizontal gradients of the velocity field *and* an energetic flow field. Figure 4.4a shows the meridional acceleration (in m/s per

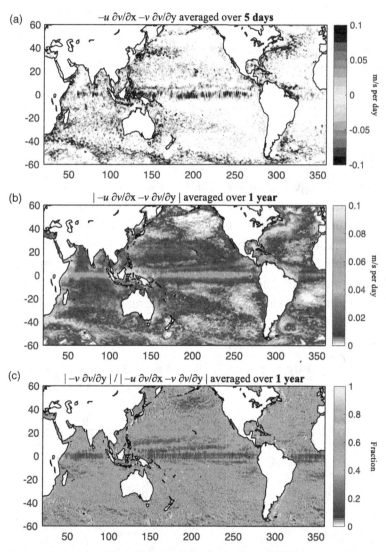

Figure 4.4 Horizontal advection of meridional velocity ($-u\,\partial v/\partial x - v\,\partial v/\partial y$) during the last pentad of 2018 from 1/3° OSCAR surface currents (a), the absolute value of $-u\,\partial v/\partial x - v\,\partial v/\partial y$ averaged over 2018 (b), and the ratio of meridional advection of meridional velocity ($-v\,\partial v/\partial y$) to the total horizontal advection of meridional velocity (c) (absolute values, averaged over 2018).(A black and white version of this figure will appear in some formats. For the color version, please refer to the plate section.)

day) due to the horizontal advection of meridional velocity, i.e., $-u\ \partial v/\partial x - v\ \partial v/\partial y$ using near-surface currents u and v averaged over a five-day period at the end of 2018. Figure 4.4b shows the absolute value of the advection term averaged over the entire year 2018, which gives a sense of where momentum advection is important in general, even for locations where its sign reverses frequently. In general, the advection term is important in three provinces: the western boundaries of each ocean basin (e.g., the Gulf Stream region off the US east coast), near the equator, and along the Antarctic Circumpolar Current (ACC) in the Southern Ocean. In the case of the Gulf Stream, for example, horizontal advection of meridional momentum should be strong because western boundary currents are fast (large v). This is confirmed in the bottom panel, which shows that the nonlinear advection term $-v\ \partial v/\partial y$ constitutes a dominant fraction of the total horizontal advection of meridional velocity. In the equatorial belt, on the other hand, meridional momentum is advected primarily by zonal velocity; this is the signature of undulating tropical instability waves (TIWs), which carry sharp zonal gradients of meridional velocity (large $\partial v/\partial x$) while propagating westward with the mean surface current ($u < 0$). The strength of the advection term along the Southern Ocean is a mixed bag, being home to perpetual, intense eddy activity amid the famously swift ACC.

4.2.5   Putting It Together and Breaking It Down

Having now accounted for all of the major processes by which seawater can gain or lose momentum, i.e., accelerate or decelerate including mere turning, it is time to combine the mathematical terms into a complete momentum budget and make plans for how we might use it to understand the ocean's inner workings, interactions with the atmosphere, and role in global climate variability.

By gathering our final expressions for the three main geophysical forces that act on a fluid (gravity [including pressure], Coriolis, and friction) as well as advection, we can write the complete set of equations of motion as

$$\frac{\partial u}{\partial t} = -\frac{1}{\rho}\frac{\partial p}{\partial x} + fv + A_h \nabla_h^2 u + A_z \frac{\partial^2 u}{\partial z^2} - u\frac{\partial u}{\partial x} - v\frac{\partial u}{\partial y} - w\frac{\partial u}{\partial z} \tag{4.19a}$$

$$\frac{\partial v}{\partial t} = -\frac{1}{\rho}\frac{\partial p}{\partial y} - fu + A_h \nabla_h^2 v + A_z \frac{\partial^2 v}{\partial z^2} - u\frac{\partial v}{\partial x} - v\frac{\partial v}{\partial y} - w\frac{\partial v}{\partial z} \tag{4.19b}$$

$$\quad\ \ \text{A} \qquad\ \ \text{B} \quad\ \text{C} \quad\ \ \text{D} \qquad\quad \text{E} \qquad\ \ \text{F} \qquad\ \text{G} \qquad\ \text{H}$$

$$\frac{\partial w}{\partial t} = -\frac{1}{\rho}\frac{\partial p}{\partial z} - g - u\frac{\partial w}{\partial x} - v\frac{\partial w}{\partial y} - w\frac{\partial w}{\partial z} \tag{4.19c}$$

$$\qquad\quad \text{I} \quad\ \ \text{J}$$

$$\frac{\partial u}{\partial x} + \frac{\partial v}{\partial y} + \frac{\partial w}{\partial z} = 0 \tag{4.19d}$$

$$\quad\ \text{K} \quad\ \ \text{L} \quad\ \ \text{M}$$

while keeping in mind that it will soon be useful to invoke the equivalence of A_z $\partial^2 u/\partial z^2$ with $1/\rho \, \partial \tau_x/\partial z$ and $A_z \, \partial^2 v/\partial z^2$ with $1/\rho \, \partial \tau_y/\partial z$ based on the definition of viscosity if wind stress is known. Recall that the continuity equation is necessarily a part of the equations of motion to ensure that we have a solvable system of equations, i.e., the same number of equations (four) as so-called unknowns (u, v, w, and p). Also recall that the Coriolis and friction terms are neglected in the vertical momentum budget. The following list summarizes the vocabulary used to define each term.

A = acceleration
B = horizontal pressure gradient force
C = Coriolis force
D = horizontal eddy diffusion
E = vertical eddy diffusion (or vertical divergence of wind stress)
F, G, H = zonal, meridional, and vertical advection of velocity
I = vertical pressure gradient force (directed upward)
J = gravity (directed downward)
K, L, M = zonal, meridional, and vertical divergence.

As we endeavor to apply these equations of motion to understanding the salient features of the global ocean circulation, there will be assumptions or approximations to be made along the way, and these usually begin with simplifying Equation 4.19 by justifying neglect of one or more of its terms. Such simplifications were first explored decades – if not centuries – before our time and have ultimately led to an impressive level of understanding and predictive capacity of the ocean circulation! Before specific applications, it is worth defining in a general sense the circumstances under which each term might be neglected.

If one is interested in the time-mean balance of forces, then one can make the familiar **steady-state** approximation by eliminating or setting the time derivative (term A) to zero. This is really a matter of time scale; rarely would one actually assume that the velocity field in the ocean is constant. It is useful for understanding what is going on at one instant in time, or the ocean circulation averaged over perhaps a year or more. This is the first and main approximation we will make in the following section. Moving through Equation 4.19, if we are not interested in accelerations due to the pressure gradient force because, for example, we are interested in attributing momentum changes driven solely by wind stress, then we could eliminate term B. Given that, as explained in Section 4.2.1, pressure gradients arise from horizontal differences in the height of the water column, an apt moniker for this assumption is **flat** (that is, the sea surface is flat). If one is interested in the dynamics of momentum very near the equator, then one could either eliminate the Coriolis force (term C) or replace it with a value that simply varies as a constant, linear function of latitude – the so-called **β-plane** approximation. If one wishes to neglect friction altogether in order to, for example, understand strictly the impact of pressure gradients and other forces on the circulation, then one might eliminate both terms D and E and logically refer to the fluid or the flow as **inviscid** (without viscosity, or at least without mixing by turbulent eddies). It is more common in practice to neglect only term D (horizontal eddy diffusion) for either suspicion of its smallness or lack of sufficient data to constrain it, while retaining term E in the form of wind stress. Finally, one might argue that the flow is sufficiently **homo-geneous**[1] or constant as a function of space (i.e., $\mathbf{V}(x, y, z) = \mathbf{V}(x_0, y_0, z_0)$), in which case there can be no advection (terms F, G, and H) as there are no spatial gradients of momentum. Making such an assumption is a matter of both spatial scale and the characteristics of the circulation. Whether making the steady state, flat, β-plane, inviscid, or homogeneous assumption, one should always do so with care – that is, articulate the justifications and of course frame the results within that context.

4.3   Geostrophic Currents: A Direct Application of the Momentum Budget

Having erected a set of rather lengthy equations thus far in this chapter, let us now take the opposite tack and try our hand at simplifying them in the interest of learning something about nature. Consider the steady-state solution to the horizontal momentum budget (Equations 4.19a and 4.19b), where the flow is also inviscid and homogeneous. Now you know that's just fancy talk for: cross-out terms A and D–H. Such approximations reduce the horizontal momentum budget to

$$0 = -\frac{1}{\rho}\frac{\partial p}{\partial x} + fv \qquad (4.20a)$$

$$0 = -\frac{1}{\rho}\frac{\partial p}{\partial y} - fu \qquad (4.20b)$$

and rearranging yields

$$\frac{1}{\rho}\frac{\partial p}{\partial x} = fv \qquad (4.21a)$$

$$\frac{1}{\rho}\frac{\partial p}{\partial y} = -fu \qquad (4.21b)$$

which is a premise that the pressure gradient force exactly balances the Coriolis force. This force balance is widely referred to across both oceanography and meteorology as **geostrophic balance**. Solving for u and v, we have

$$v_g = \frac{1}{\rho f}\frac{\partial p}{\partial x} \qquad (4.22a)$$

---

[1] Some authors and researchers use the term homogeneous to describe the context in which not only advection but also the pressure gradient force and even viscosity are neglected.

$$u_g = -\frac{1}{\rho f}\frac{\partial p}{\partial y} \tag{4.22b}$$

where the subscript g is introduced to serve as a reminder that these velocities carry the implicit assumption that geostrophic balance holds (and all of the other terms of the horizontal momentum budget that were neglected in deriving them are still neglected). To interpret Equation 4.22, a positive meridional (northward) geostrophic velocity would be induced by a positive zonal pressure gradient, and a positive zonal (eastward) geostrophic velocity would be induced by a negative meridional pressure gradient, in the Northern Hemisphere where $f > 0$.

Let's take a step back and notice what just happened, as there was no sleight of hand. Our first step in this derivation was to eliminate the velocity tendencies that are usually the object of our interest (that is, $\partial u/\partial t$ and $\partial v/\partial t$), but because the Coriolis force in the zonal momentum budget is a function of meridional velocity and vice versa, we were able to derive a simple expression for both velocity components (with appropriate assumptions) that *only* requires us to know about the horizontal pressure gradient (f is a trivial function of latitude)! Meteorologists have little difficulty at this point, since we humans live at the bottom of the ocean of air where horizontal pressure gradients are substantial and resultant wind currents are interesting. Moreover, civilization has spread sufficiently across Earth's surface that weather balloons launched several times daily carrying inexpensive barometers can characterize the full three-dimensional pressure field of the entire troposphere and lower stratosphere with enough precision to calculate the geostrophic wind everywhere and even compare it to independently measured winds. If only humans also had a way to measure pressure routinely throughout the world's oceans with such precision to reliably compute its gradients and plug them into Equation 4.22.

But what if there was a different form of the geostrophic velocity equations that we *could* solve using some presently observable quantities in the ocean? Beginning with the vertical momentum budget and applying the same approximations as above (steady state, inviscid, and homogeneous), Equation 4.19c reduces to

$$\frac{1}{\rho}\frac{\partial p}{\partial z} = -g \tag{4.23}$$

This equilibrium condition in which the upward vertical pressure gradient force is balanced by the downward force of gravity is appropriately named **hydrostatic balance** (*static*, as in the water is not moving – at least vertically – since there is no appearance of w in Equation 4.23). If we multiply through by density ρ, we obtain

$$\frac{\partial p}{\partial z} = -\rho g \tag{4.24}$$

and integrating both sides with respect to depth z yields an expression for pressure, referred to in this case as **hydrostatic pressure**, which is simply a function of depth (as ρ and g can effectively be considered constants for now):

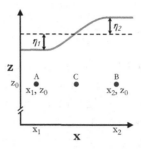

Figure 4.5 Schematic diagram of a sloping sea surface (gray line) relative to the geoid (dashed black line), defining the variable sea surface height (η) and illustrating that horizontal gradients in sea surface height give rise to (and are sufficient to characterize) horizontal pressure gradients in the ocean.

$$p = \rho g z \qquad (4.25)$$

If pressure is only a function of depth, then how can there be a horizontal pressure *gradient* between two adjacent locations in the ocean?

The answer is that the sea surface is not flat (nor did we assume it to be so in our derivation of geostrophic currents), so two otherwise adjacent locations at the same depth can actually be at slightly different distances below the sea surface and therefore different depths. Such a scenario is depicted schematically in Figure 4.5. If the sea surface were flat, as would nearly be the case if the ocean was completely at rest, then points A and B would have the exact same depth of water resting above them and therefore the zonal pressure gradient $\partial p/\partial x$ at point C would be zero. As an aside, this hypothetical "flat" sea surface depicted by the dashed line in Figure 4.5 is called the **geoid** and is now known reasonably well thanks to the field of geodesy, which includes the study of Earth's shape and gravitational field. It is not quite flat, though, because features of the seafloor from broad mid-ocean ridges to petite seamounts exert a gravitational pull and cast reflections of themselves even on a motionless sea surface. In the more typical case where the sea surface is not flat relative to the geoid, such as that depicted by the gray line in Figure 4.5, p as calculated from hydrostatic pressure (Equation 4.25) is clearly different at points A and B such that $\partial p/\partial x > 0$ at point C and, from Equation 4.22a, implies a positive (northward) geostrophic meridional velocity v_g there (i.e., directed into the page) assuming the diagram represents a scenario in the Northern Hemisphere. However, as it is only the *difference* in pressure between adjacent points that matters for geostrophic velocity, and the main source of such a difference is a difference in the sea surface height above adjacent points, we need only knowledge of the sea surface height field (denoted by η). In other words, and very conveniently, $\Delta\eta \propto \Delta p$.

For this reason, and in recognition of the importance of the ocean circulation in general, substantial international investments and remarkable engineering progress has been made in developing our capacity to measure sea surface height relative to the geoid frequently, globally, and precisely. It is the least we can do here, knowing it is bound for a horizontal partial derivative or finite difference anyway, to replace z in Equation 4.25 with η and substitute the resulting expression ($\rho g \eta$) for p in Equation 4.22, by which we obtain geostrophic velocity as a function of the sea surface height gradient:

$$v_g = \frac{g}{f}\frac{\partial \eta}{\partial x} \tag{4.26a}$$

$$u_g = -\frac{g}{f}\frac{\partial \eta}{\partial y} \tag{4.26b}$$

A simple application of the geostrophic velocity equations to the observed sea surface height field is shown in Figure 4.6. The observed sea surface height field, in this case, is the time-mean as observed by a series of satellite altimeters

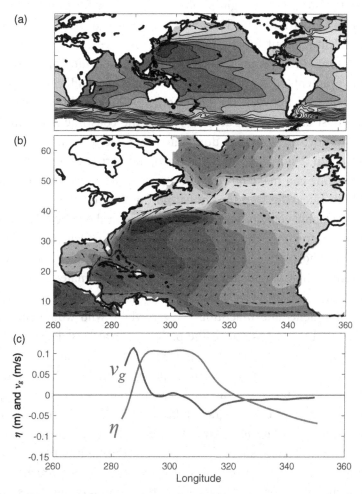

Figure 4.6 Sea surface height above geoid η averaged from 1993 to 2010 from the 1° AVISO satellite altimeter product with contour interval 0.2 m and values spanning roughly −1.5 to 1.5 m (a). Closer view of η and calculated geostrophic velocity vectors in the North Atlantic region with η contour interval 0.1 m and values spanning roughly −0.6 to 0.9 m (b). Longitudinal profiles of η and v_g along 36.5°N latitude (c). To align the two profiles, η was plotted as $0.3\eta - 0.1$. (A black and white version of this figure will appear in some formats. For the color version, please refer to the plate section.)

between 1993 and 2010. An altimeter is an instrument carried by Earth-orbiting satellites that measures its own height above the sea surface very precisely (centimeter precision) by carefully measuring the time it takes a pulse of microwave energy emitted by the satellite to reach the surface, reflect, and return to the altimeter. By doing so continuously, as the satellite orbits from pole to pole with the planet spinning beneath it, near-global coverage is achieved in a few days and by this method we now have almost three decades of global, high-resolution (in both space and time), and precise observations of η, the sea surface height above geoid. While it may seem as though an altimeter is only able to provide information about the surface of the ocean, as you will see in later chapters, it actually tells us a great deal about the subsurface thermal structure of the ocean as well.

Let us, just for now, take the global distribution of η for granted (Figure 4.6a), and simply apply Equation 4.26. The result for the North Atlantic region is shown in Figure 4.6b. Throughout the domain, geostrophic velocity vectors are running parallel to contours of constant sea surface height, and they are strongest where the horizontal gradients of η are sharp. For example, the Gulf Stream is noticeably strong, and is aligned between the high point in the η field well east of the US mid-Atlantic coast and where it drops down to the coastline itself. In the central and eastern Atlantic, the η field is dominated by a weak negative zonal gradient, which brings about a slow southward geostrophic current. Looking closely at the Gulf of Mexico, one can make out the Loop Current flowing along contours lining a bubble of high η leaking into the Gulf from the Caribbean Sea.

That the geostrophic velocity vectors align well with the η field is not surprising; in fact, it is by construction. The author's computer simply calculated $\nabla\eta$, multiplied the result by g/f and plotted it on top of η. However, students of oceanography needn't even go to that length just to derive a quick estimate, even if qualitative, of the geostrophic flow provided by a map of η. Presenting the ball trick, which is a favorite of students and practitioners everywhere who would find it desirable to leave differential equations out of the picture – or at least feel like they are. To infer the geostrophic velocity at some spot on a map of η, move your eye a short distance "uphill" from that spot (i.e., to a spot with higher η) and imagine placing a ball there. Since pressure gradients are the source of energy for geostrophic flow, which is ultimately due to gravity (as explained in Section 4.2.1), the ball will obviously roll down the hill. Then, to mimic the Coriolis force, turn the ball to the right (or left, if your spot of interest is in the Southern Hemisphere) until the ball is traveling parallel to the lines of constant η. There you have your geostrophic flow, and the more densely packed those topographic lines, you can be sure the faster the geostrophic flow. This simple trick can also be used to double-check your mathematically derived work, of course. An illustration of the systematic dependence of geostrophic velocity on η is shown in Figure 4.6c, in the form of longitudinal profiles of η and v_g along latitude 36.5°N. Working our way eastward from the US east coast, there is a sharp rise in η, a plateau, followed by a decline that

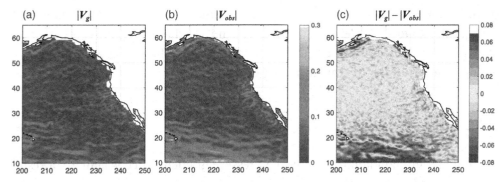

Figure 4.7 Near-surface currents in 2012 as calculated by applying the geostrophic velocity equations to observed sea surface height (a), as estimated by the OSCAR satellite product (b), and their difference (c). The units in all panels are m/s. (A black and white version of this figure will appear in some formats. For the color version, please refer to the plate section.)

becomes ever more gradual all the way to the eastern boundary of the Atlantic, where Africa meets Europe. This pattern clearly explains the strong spike in northward velocity that we call the Gulf Stream and the broad, slow southward flow spanning most of the rest of the way across the ocean – a classic juxtaposition of western and eastern boundary currents that we will explore in more depth in subsequent chapters.

How closely do geostrophic currents match the real currents of the ocean? Surely, we have neglected some major contributors to the ocean's circulation here in our investigation of geostrophic flow, including (but not entirely) wind and density-driven momentum. (The ominous "not entirely" is meant to remind you that we simply took the global distribution of sea surface height for granted, bothering not with why it looks the way it does.) A comparison between geostrophic currents as calculated by Equation 4.26 and a more comprehensive estimate of the actual near-surface currents (accounting not only for geostrophy but also wind and temperature effects) is portrayed for the northeastern Pacific Ocean (Figure 4.7). The currents in both cases represent the average scalar speed over the year 2012. At the relatively finer scale, most of the features in the region are indistinguishable, suggesting that sea surface height and therefore geostrophy is an indispensable diagnostic tool! Where they differ substantially (Figure 4.7c) is in the tropics where, between about 10°N and 20°N, the geostrophic calculation underestimates the speed by nearly 50 percent, highlighting the need to return and take aim at some of the remaining physics that we neglected by way of simplifying the momentum budget at the beginning of our derivation of geostrophic flow.

Further Reading

Many examples of ocean momentum budget analyses for the equatorial oceans are available in the scientific literature, including Wacongne (1989) for the equatorial Atlantic Ocean and Nagura and McPhaden (2008, 2014) for the equatorial Indian

Ocean. Studies focusing on the equatorial Pacific Ocean include McPhaden and Taft (1988), Qiao and Weisberg (1997), Yu and McPhaden (1999), Brown *et al.* (2007) and Drenkard and Karnauskas (2014). These studies are especially illuminating in terms of the treatment of friction and wind stress (Section 4.2.3).

Questions

1. Explain in plain language why the pressure gradient force is actually a result of gravity.
2. The β-plane approximation simplifies some ocean dynamics. In plain language, why is it far more defensible near the equator than at high latitudes?
3. What are the key assumptions or approximations that one makes in order to reduce the horizontal momentum budget to an expression of geostrophic balance? What further approximation can one leverage to express geostrophic balance in terms of sea surface height?
4. Why is it practically desirable for physical oceanographers to obtain an expression of geostrophic velocity as a function of sea surface height rather than pressure?
5. Examine the following latitude-by-depth cross-section, illustrating the sea surface height (η) and zonal geostrophic velocity (u_g), where $u_g > 0$ (i.e., eastward, or pointing out of the page). Which hemisphere is this diagram valid for?

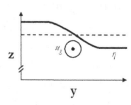

6. In the above diagram, say $u_g = 0.5$ m/s and it is valid for somewhere right on the edge of the tropics ($\pm 23.5°$). What is the slope of the sea surface, i.e., the local meridional gradient of sea surface height ($\partial\eta/\partial y$) in units of millimeters per kilometer, immediately above the eastward geostrophic current featured in the diagram?
7. Briefly explain why the sea surface height field η is a mirror image of the thermocline depth.
8. Consider a so-called "cold core ring" that has broken away from the Gulf Stream in the North Atlantic Ocean, as depicted below. Due to the reduced amount of heat contained in the near-surface waters inside of this swirling eddy, it is associated with a lower sea surface height than its surroundings. (a) Explicitly state the sign (positive, negative, or zero) of both of the geostrophic velocity components (u_g and v_g) at points A, B, and C. (b) Assuming the cold

core ring is indeed in geostrophic balance, how would you expect the overall surface circulation in the vicinity of this ring to look (e.g., clockwise, counterclockwise, inward toward center, outward from center, no motion at all)?

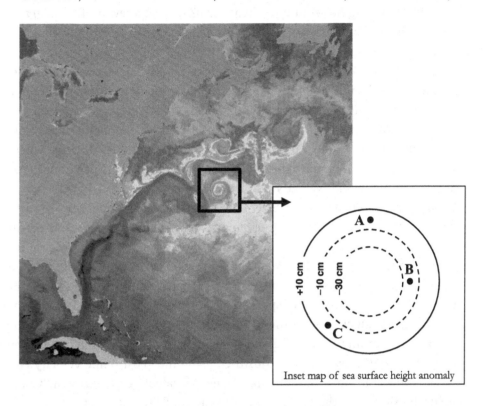

Inset map of sea surface height anomaly

9. Sketch an ocean surface map, valid for the Southern Hemisphere, featuring isohalines and sea surface height contours such that the sea surface salinity would increase over time at a point labeled on the map due to geostrophic salinity advection. You may assume $E = P$ at the point you specify. It is recommended to use solid lines for isohalines and dashed lines for η contours. The diagram must be labeled completely, including a clear indication of which contour lines have relatively higher and lower values of SSS and η.

10. A recent study analyzed coastal tide gauge records to reveal that sea levels along the US east coast, north of Cape Hatteras, have been rising by about 4 mm per year over recent decades, which is 3–4 times faster than the global average rate and also greater than those further offshore of the mid-Atlantic US coast. Based on this finding, would you expect that the Gulf Stream north of Cape Hatteras has been strengthening or weakening in recent decades?

11. Select one of the *Dive into the Data* boxes featured in this chapter. Read the sample journal article that uses the associated data set, and describe the general role of that data set in the study. Provide a little context, both technical and scientific. What did the authors actually "do" with the data set? What specific scientific insight was enabled by incorporating this data set into the study?

12. Utilizing *Dive into the Data* Box 4.3, produce a global map of and examine the time-averaged sea surface height field. (a) In broad-brush terms, where in the world ocean is η relatively high and relatively low? (b) Where are there particularly strong horizontal gradients of η? (c) Plot a few profiles of η along various lines of latitude; are zonal η gradients stronger along the equator or in the midlatitudes ($\sim 30°$N)?

13. Following on from part (c) of the previous question, calculate and plot a longitudinal profile of the meridional geostrophic velocity $v_g(x)$ somewhere in the midlatitudes using the η observations. Then, using surface current observations from *Dive into the Data* Box 4.2, overlay a profile of meridional velocity $v(x)$ at the same latitude. Zoom in to a particular longitude. Describe whether your graphs aptly illustrate the functional relationship between zonal η gradients and meridional velocity and why.

14. Utilizing *Dive into the Data* Box 4.3, using the *high-resolution* sea surface height field, calculate the time-averaged geostrophic velocity and produce global maps of u_g, v_g, and the scalar geostrophic speed $\sqrt{u_g^2 + v_g^2}$. Zoom in to the ocean region just east of Australia. In a 2003 Pixar film, Marlin and Dory ride the East Australian Current (EAC) to "find Nemo." (a) Identify the EAC on your map of geostrophic speed. Without regard to direction (yet), which major Australian cities does the EAC connect? (b) Strictly based on the global η map (or a profile of η along some line of latitude), what direction do you predict the EAC flows and why (justified using the geostrophic velocity equations)? Now confirm your prediction using the calculated geostrophic velocity fields. (c) How does the EAC compare to the Gulf Stream and the Kuroshio in terms of geostrophic speed? (d) How large and sharp of a zonal η gradient is associated with the EAC (provide a quantitative estimate in meters) and, qualitatively, how does this compare to the zonal η gradients associated with the Gulf Stream and Kuroshio Currents? (e) If Marlin and Dory were texting while swimming and neglected to exit the EAC at the second major Australian city linked by the EAC, where would they end up, and approximately how long would it take to get there?

DIVE INTO THE DATA BOX 4.1

Name   Cross-Calibrated Multi-Platform (CCMP) Ocean Surface Winds
Synopsis Blend of satellite estimates (over a dozen satellites merged) and *in-situ* (anemometers on buoys) observations of near-surface (10 m) winds over the ocean. Such measurements from space are made possible by instruments called scatterometers (and others).

DIVE INTO THE DATA BOX 4.1 (cont.)

Scatterometers detect the amount of "scatter" of reflected radiation off the ocean surface, which is related to the roughness of the sea surface, which, in turn, is related to the wind speed. Such a data set is extremely valuable since sustained observations of surface winds away from land, ships, and buoys where anemometers can be placed would otherwise be impossible to obtain.

Science Since wind plays a major role in driving the ocean circulation, the CCMP data set enables investigation of ocean–atmosphere interaction at relatively high resolution and across temporal scales from days to decades. Wind components (zonal and meridional) can be easily converted to wind stress (or pseudostress), which can be used to calculate Ekman transport, and derivatives of the wind stress field such as divergence and curl can be computed to diagnose features of the wind-driven ocean circulation such as upwelling.

Figures 4.3, 5.10, 5.11, 5.12, 6.3, 6.7

Version 2, Level 3.5*

Variable Surface wind velocity

Platform Satellite and mooring

Spatial Global (gridded), 0.25° resolution

Temporal 1988–2017, monthly averages**

Source http://data.remss.com/ccmp/v02.0

Format NetCDF (.nc)

Resource ccmp.m; ccmp.mat (sample data and code provided on publisher website here: www.cambridge.org/karnauskas)

Journal Reference for Data

Atlas, R., Hoffamn, R., Ardizzone, J., *et al.* A cross-calibrated, multiplatform ocean surface wind velocity product for meteorological and oceanographic applications. *Bull. Amer. Meteor. Soc.* **92**, 157–174 (2011). DOI: 10.1175/2010BAMS2946.1

Sample Journal Article Using Data

Carranza, M. M. and Gille, S. T. Southern Ocean wind-driven entrainment enhances satellite chlorophyll-a through the summer. *J. Geophys. Res. Oceans* **120**, 304–323 (2015). DOI: 10.1002/2014JC010203.

* In 2011/2012, Remote Sensing Systems (RSS) took over maintenance, update, and continued processing of CCMP, including a full reprocessing of the original data (up to 2011). This is referred to as CCMP v.2. The original data set produced by NASA is still available here: ftp://podaac-ftp.jpl.nasa.gov/allData/ccmp/L3.5a with essential information here: https://podaac.jpl.nasa.gov/dataset/CCMP_MEASURES_ATLAS_L4_OW_L3_5A_MONTHLY_WIND_VECTORS_FLK.

** Six-hourly resolution also available (Level 3.0).

DIVE INTO THE DATA BOX 4.2

Name   Ocean Surface Current Analysis Realtime (OSCAR)

Synopsis Like surface winds over the ocean, sustained observations of surface currents are challenging. Unlike winds, however, direct estimates from satellites are not presently feasible. The OSCAR data set therefore calculates the near-surface currents over the global ocean using core concepts of physical oceanography, including geostrophy and Ekman dynamics. For example, satellite observations of sea surface height are used to calculate the geostrophic component of the flow, and satellite observations of surface winds are used to calculate the Ekman component of the flow. The data represent the currents averaged over the upper 30 m of the ocean.

Science   Near-surface ocean currents are used in real-time ocean forecasting applications as well as scientific investigations of the ocean circulation. The relatively high resolution enables study of eddies, and the global domain (and nearing 30 years of temporal coverage) of the data set enables studies of the larger-scale ocean circulation and climate.

Figures   2.8, 4.4, 4.7, 6.7

Version   1, Level 4

Variable Near-surface current velocity

Platform Satellite

Spatial   Nearly global (66°S–66°N) (gridded), 0.33° resolution*

Temporal 1993 to present, five-day averages

Source   ftp://podaac-ftp.jpl.nasa.gov/allData/oscar/preview/L4/oscar_third_deg

Format   NetCDF (.nc)

Resource oscar.m; oscar.mat (sample data and code provided on publisher website here: www.cambridge.org/karnauskas)

Journal Reference for Data

Bonjean, F. and Lagerloef, G. S. E. Diagnostic model and analysis of the surface currents in the tropical Pacific Ocean. *J. Phys. Oceanogr.* **32**, 2938–2954 (2002). DOI: 10.1175/1520-0485(2002)032<2938:DMAAOT>2.0.CO;2.

Sample Journal Article Using Data

Foltz, G. R. and McPhaden, M. J. Seasonal mixed layer salinity balance of the tropical North Atlantic Ocean. *J. Geophys. Res.* **113**, C02013 (2008). DOI: 10.1029/2007JC004178.

* Also available at 1° resolution.

DIVE INTO THE DATA BOX 4.3

Name   AVISO Sea Surface Height

Synopsis   This sea surface height data set from Archiving, Validation and Interpretation of Satellite Oceanographic data (AVISO) blends several satellite altimeters to form a consistent estimate of absolute dynamic topography. Absolute dynamic topography is defined as the sea surface height relative to the geoid. The geoid is the theoretical ocean surface if it were at rest, which would not be perfectly flat due to the gravitational pull of seafloor features.

Science   Sea surface height observations can be used to compute geostrophic currents and to study modern sea level rise due to anthropogenic greenhouse gas emissions. While tide gauges have been in place and measuring sea level *in situ* for much longer than altimeters have been in orbit, they are confined to coastal sites, whereas satellite altimeters provide a global view and key forces acting on the ocean such as the pressure gradient force, which depends on the spatial gradients of sea surface height, can be computed.

Figures   4.6, 4.7, 6.7

Version   Level 4

Variable   Absolute dynamic topography

Platform   Satellite

Spatial   Global (gridded), 1° resolution*

Temporal   1993–2010, monthly averages

Source   ftp://podaac-ftp.jpl.nasa.gov/allData/aviso/L4/dynamic_topo_1deg_1mo

Format   NetCDF (.nc)

Resource   aviso.m; aviso.mat (sample data and code provided on publisher website here: www.cambridge.org/karnauskas)

Journal Reference for Data

N/A, see www.aviso.altimetry.fr/en.

Sample Journal Article Using Data

Dong, S., Gille, S. T., and Sprintall, J. An assessment of the Southern Ocean mixed layer heat budget. *J. Climate*, **20**, 4425–4442 (2007). DOI: 10.1175/JCLI4259.1.

* Also available at 0.25° resolution.

5 | The Atmospheric Interface

5.1 Motivation and Scope of this Chapter

The previous three chapters left the door open for a strong control of the atmosphere on the ocean's mixed layer heat and salt content, and its circulation via the momentum budget. For example, some of the processes comprising the net surface heat flux term (Q_{net}) of the ocean mixed layer heat budget (Chapter 2) are directly influenced by the speed and thermodynamic state (T and q) of the atmosphere where it meets the ocean – the turbulent fluxes. Others depend on the abundance of shortwave radiation reaching the surface and the amount of longwave radiation absorbed on its way out to space – the radiative fluxes – and those, in turn, depend on the composition of the atmosphere above, including clouds, water vapor, and even ozone. We also had a preview of how the atmosphere has inroads to the ocean's momentum budget (Chapter 4) via friction in the form of wind stress (τ), which will be expanded upon in the next chapter to ascertain one of the two major components of the global ocean circulation. It is one of the aims of this chapter to put faces to those names.

In the physical oceanography of ages ago, it was customary to treat those atmospheric variables and processes simply as "inputs" to the ocean, but the climate dynamics of today require a healthy appreciation of the various mechanisms by which the two fluids interact. This is where the field is heading in terms of understanding, and certainly the principles on which state-of-the-art numerical models of the climate system are built. While the point of contact may be at the surface, the physics governing such interactions extend upward and downward throughout virtually the entire depths of each fluid. Consider the following with your imagination. The air and sea are but one stratified fluid and it is routine to probe its temperature – from top to bottom – at a single geographic location. What would a *single* temperature profile look like, if on the same vertical scale? Try to sketch one before turning the page. It turns out that it is possible to construct such a thing where oceanographic moorings are deployed near small islands from which weather balloons are launched (Figure 5.1). That there must be a *continuous* temperature profile $T(z)$ is common sense, but the depth range over which negotiations play out for the surface climate is fun to envision from this perspective. In departing from the canonical vertical coordinate in the atmospheric sciences (pressure), it is also illuminating to see the incredible sharpness of the ocean's vertical stratification compared to that of the atmosphere. Whereas a 20 °C drop in temperature from that of the surface occurs over ~3000 m in the atmosphere, the same 20 °C drop is achieved in just the first 300 m of the ocean! With such a sharp vertical gradient of temperature $\partial T/\partial z$ so close to the surface, following Chapter 2, one can infer that it wouldn't take much of a vertical velocity w to provoke a substantial change in the sea surface temperature (SST) via advection ($-w\,\partial T/\partial z$), with which the atmosphere will be forced to, somehow, equilibrate.

This chapter is not a crash course in atmospheric dynamics and thermodynamics. Whether you have no prior knowledge of the atmospheric or you have a solid, complementary background in atmospheric science and wish to make a more deliberate consideration of the ocean in your studies or researches on climate, this chapter provides some essentials of the large-scale structure and circulation of the atmosphere needed to understand *what determines the state of the atmosphere along its interface with the ocean* and, in turn, *the ways in which the atmosphere is sensitive to the state of the ocean.*

We begin with a broad view of the structure of the atmosphere, the global energy balance (and imbalance) and an intuitive piecing together of the emergent

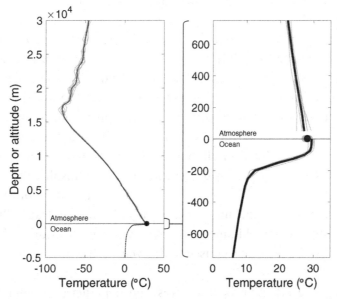

Figure 5.1 Merged measurements of temperature in the atmosphere and ocean in the western tropical Pacific on July 24, 2010. The figure includes vertical profiles of atmospheric temperature from weather balloons (data spanning 50 m to 30 km, i.e., well into the stratosphere), moored observations of near-surface air temperature (at 2.2 m), and vertical profiles of subsurface ocean temperature from the same moorings (data spanning −1.5 m to −750 m; the dotted line is extrapolation for illustrative purposes). The right panel zooms into the surface ±750 m. Gray lines and circles are individual profiles, while black lines and circles are the average of all profiles on that day.

general circulation, followed by some specific mechanisms by which the atmosphere responds to oceanic forcing, including SST anomalies. In Chapter 6 we will return to the ocean's behavior in the same vein and discuss some specific mechanisms by which the ocean responds to atmospheric forcing; in Chapter 7 these themes will be united in an understanding of the two-way *coupled* interactions between the ocean and atmosphere that give rise to the rich variety of climate variability on Earth.

5.2   The Vertical Structure of the Atmosphere

Virtually every survey course in weather, climate, and many other branches of the Earth sciences demands that students memorize the names and defining characteristics of the layers of Earth's atmosphere. Let us not miss such a thrill here, but more importantly, let's couple the occasion with some simple, physical rationalization for their existence. There are four main layers of the atmosphere as defined by temperature, but not according to whether they are warm or cold layers; rather, on the sign of the vertical temperature gradient $\partial T/\partial z$. Why this is so is fairly obvious in hindsight; most of the atmosphere from the surface to space indeed *has* an appreciable vertical temperature gradient rather than being even close to isothermal.

Consider a planet wrapped in a simple atmosphere that is entirely transparent to shortwave (or solar) radiation. So long as that planet has a surface temperature, it will emit longwave radiation and the air temperature will decrease more-or-less linearly[1] with distance from the heat source (Figure 5.2). This is the starting point of our more comprehensive, top-to-bottom atmospheric temperature profile. Let's relax the assumption that the atmosphere is entirely transparent to solar radiation and allow the air closest to the Sun to absorb some of its high-energy radiation. This process warms the atmosphere from the top down, as illustrated in step 2 of Figure 5.2. To complete our sketch of the atmosphere, let's introduce a third (and final) warming agent – ozone. Naturally occurring ozone (O_3) residing tens of kilometers above the surface absorbs ultraviolet radiation emitted by the Sun and consequently warms the air, as illustrated in step 3 of Figure 5.2.

With three effective heat sources and a perpetual tendency to lose heat in between, there are four easily distinguished layers based on $\partial T/\partial z$. These layers have been named, from the bottom up, the **troposphere**, **stratosphere**,

---

[1] Depending on the composition and density structure of the hypothetical atmosphere, the change in temperature with height may just as well be *nonlinear* as implied by the inverse square law, a distinction with no bearing on the result of this thought experiment.

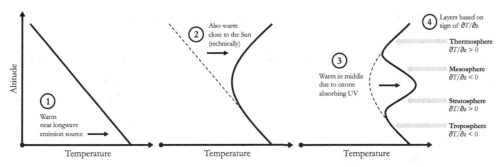

Figure 5.2 Idealized vertical structure of the atmosphere (in terms of the vertical profile of the temperature) in three simple steps. In each step, the temperature profile resulting from the previous step is presented as a dashed line. Temperature units are arbitrary and the plots are not to scale with respect to altitude.

mesosphere, and **thermosphere**, and each of the prefixes appearing before "sphere" derive from Greek or Latin words that made logical sense to notable European meteorologists at the turn of the twentieth century, such as Léon Philippe Teisserenc de Bort (1855–1913) and Richard Assman (1845–1918). In the real-world vertical temperature profile shown in the previous section, one can see that the boundary between the troposphere and stratosphere is about 17 km in the western tropical Pacific region. This boundary is known as the **tropopause**, and each boundary is so named with the suffix "pause" following the prefix associated with the *lower* layer (so the stratopause marks the boundary between the stratosphere and mesosphere, and so on). The tropical tropopause, especially that in the western tropical Pacific, happens to be much higher than the tropopause near the poles, where it is only around 10 km. Reasons for both the latitudinal and longitudinal variations in tropopause height (or, equivalently, tropospheric depth), will be made clear later in this chapter. For now, think ideal gas law – warmer tropical air occupies greater volume. The tropopause plays an important role in setting the atmospheric circulation. Even thunderstorms with intense updrafts have difficulty penetrating the tropopause due to the strong, stable stratification in the stratosphere (with temperature increasing with height, i.e., an inversion). We will therefore invoke it as something of a flexible lid in a later section.

The layers above the tropopause each have their own interesting dynamics and weather/climate phenomena, which are beyond the scope of this book. In particular, steady research progress has been made in recent decades linking weather and climate variability in the stratosphere to that in the troposphere and, in turn, elucidating connections even between the stratosphere and the ocean. The following section describes the energy budget and geographic imbalance thereof for the atmosphere as a whole, but we will then focus on the resulting circulation within the *troposphere* to ascertain the wind field at the surface.

5.3   The Global Energy Balance and Its Geographic Variation

Just beyond the very top of its atmosphere (TOA), before any chance for reflection or absorption, Earth receives from the Sun an average of 341 watts of shortwave radiation per square meter of surface area (W/m^2). A wealth of satellite observations sustained over the past couple of decades has enabled us to track those 341 watts along their journey into the atmosphere toward the surface and, ultimately, back out to space. If the planet is in equilibrium, it should also be emitting a full 341 W/m^2 back to space. So iconic are the illustrations detailing the partitioning of these energy flows that they are often referred to as "Trenberth" diagrams (even when applied to other planets!) (Trenberth *et al.*, 2009). With no slight intended to their inventor, Kevin Trenberth of the National Center for Atmospheric Research in Boulder, Colorado, we will forego the illustration and simplify these numbers a bit further than usual to get to the heart of the matter – the net global energy imbalance and the transports of heat carried by the atmosphere and ocean in response.

Of the 341 W/m^2 arriving at TOA, about 30 percent is reflected back to space immediately (20 percent by the atmosphere, including clouds, and 10 percent by the surface). The other 70 percent is absorbed by Earth (20 percent by the atmosphere and 50 percent by the surface – the ocean and land). Now, with a particular interest in the ocean, which constitutes the majority of the *surface* of Earth, let's turn our attention to the fate of that 50 percent of shortwave radiation that is absorbed by the surface. How does the ocean then rid itself of those ~170 W/m^2 of energy? Recall all of the surface fluxes comprising the Q_{net} term in the ocean mixed layer heat budget (Chapter 2) that could be in play. Globally speaking, 20 percent is emitted as net longwave radiation (Q_{LW}) and the other 30 percent is shed by turbulent heat fluxes (20 percent by latent heat flux Q_{LH} and 10 percent by sensible heat flux Q_{SH}). From the physics and global maps of these terms as presented in Chapter 2, we know that for the ocean, the split between latent and sensible heat fluxes is more heavily weighted toward latent heat flux than the global average (including land) of 20 percent/10 percent.

It needs to be borne in mind that the above description of the journey of energy incident upon TOA through the climate system and back out again represents the spatial average over the entire planet. So, while the numbers quoted above would be accurate if the entire Earth was distilled into a single, representative square meter of surface area (and hence the total outgoing energy approximately balances the total incoming energy at TOA), they are *not* valid for just about any individual location on Earth (where there is usually a net TOA energy *imbalance*). At the ocean surface, for example, a disproportionate amount of turbulent heat flux out of the ocean occurs over the relatively narrow, poleward-flowing currents found at the western side of the major subtropical ocean basins (Chapter 2). Even the numbers for the TOA, especially the initial value of the incoming shortwave radiation (341 W/m²), are *highly* dependent on latitude. This fact is illustrated using satellite measurements of TOA energy fluxes taken over one recent decade (Figure 5.3). The incoming shortwave radiation at TOA is in fact strictly a function of latitude, with the annual average ranging from ~170 W/m² at the poles to just over 400 W/m² near the equator. The geographic variation of total outgoing energy at TOA is more complex, since it represents the sum of reflected shortwave and emitted longwave radiation that actually escapes the atmosphere (notice the bright, hot surfaces of the Sahara and Arabian Peninsula, which reflect a large amount of the incident shortwave radiation *and* emit a large amount of longwave radiation). Compared to the incoming radiation, the outgoing radiation is more homogeneous, with a narrower range of values – from just under 300 to about 350 W/m². The map of *net* radiation at TOA shows some zonal variation due to that of outgoing

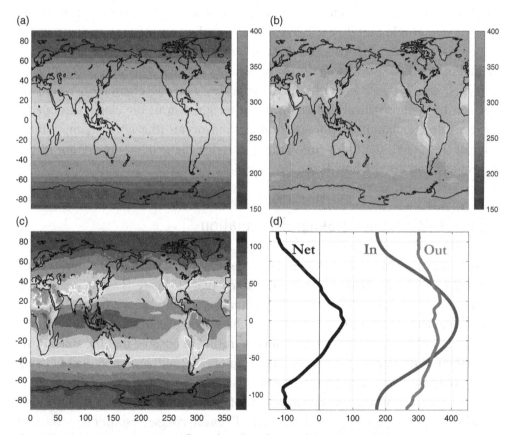

Figure 5.3 Long-term mean energy fluxes (W/m²) at the top of the atmosphere (TOA). Incoming shortwave radiation (a), total outgoing radiation (reflected solar plus longwave) (b), net radiation (incoming minus outgoing) (c), and zonal means of each quantity (d). The zero net radiation contour is shown in white in (c). Energy fluxes averaged from July 2005 through June 2015 from the NASA Clouds and Earth's Radiation Energy System (CERES) data set. (A black and white version of this figure will appear in some formats. For the color version, please refer to the plate section.)

radiation, but overall has a stronger meridional dependence such that Earth gains energy between about $\pm 35°$ latitude and overall loses energy through the TOA poleward of $\pm 35°$ latitude – by as much as $125\,W/m^2$ in the Arctic (and even more so in its dark winter).

If, overall, the high latitudes are perpetually losing heat to space and the tropics are gaining heat, why are the temperatures of these regions relatively stable over time? If the heat being gained is greater than the heat being lost in the tropics, why are they not perpetually warming? The answer is that there are *circulations* in both the atmosphere and ocean such that the excess incoming energy in the tropics is transported poleward in both directions, which serves to balance the net heat loss at those higher latitudes. Given the large disparity in heat capacities between the atmosphere and ocean, it is perhaps surprising that such a dominant fraction of this poleward heat transport is in fact performed by the atmosphere. Curious is the association between the atmosphere's circulation and the planet's latitude-dependent energy imbalance from which it emerges. Any weather system from tropical storms to the common midlatitude cyclone can be viewed as an attempt to move warm air poleward and/or cold air equatorward, but the meridional gradient of energy (including heat) is also a primary energy source of such systems. Without such gradients, weather and even the broadest scale "general" circulation would be unnecessary! Philosophical matters aside, in recognition of the close relationship between the circulations of the atmosphere and ocean (the latter being a focal point of this book), the following sections describe the general circulation of the atmosphere and discuss how it can be altered by the ocean.

5.4   The Atmospheric General Circulation

What exactly is meant by the atmospheric **general circulation** or, for that matter, climate? Mark Twain is said to have remarked that "Climate is what we expect; weather is what we get." This implies that climate is something of a faithful average of the weather, from which one might conclude that climate is just a motionless backdrop. Nothing could be further from the truth! We observe that the state of the atmosphere evolves over time scales from minutes to millennia and well beyond, and there is hardly a sharp line to be drawn separating weather and climate.

Consider the illustration of this spectrum of variability in Figure 5.4. More specifically, stare at the maps for a moment. Even at coarse resolution, a map of cloudiness averaged over a 24-hour period is sharp enough that one can quickly spot familiar weather systems such as the comma-shaped midlatitude storms parading across the oceans, masses of clouds over Scandinavia and the central USA, and even Hurricane Sandy making landfall in the northeastern USA. At an averaging interval of one week, some features that were sharp in the daily map start to appear smudged, such as those midlatitude cyclones and higher-latitude cloud masses. By one month and beyond, they are entirely blurred into the grayness of the latitude band in which they reside. This effect occurs because some such features are *moving* (primarily from west to east), like trying to take a sharp photo of a train speeding by at close range. Now let's look elsewhere – say, the deep tropics. In the daily average, there appears to be an orderly, albeit broken, band of clouds encircling the entire planet near (but just north of) the equator. These clouds arguably constituted the tropical "weather" on October 29, 2012, and yet, moving down the images all the way to the 10-year average, this feature only becomes more coherent instead of blurry like before. Indeed, that feature has a name enshrined in the canon of *climatology* – the Intertropical Convergence Zone or ITCZ. Those clouds are moving, too, but more slowly and with less space between (and to the west). Hence, we arrive at a paradox of terms: If weather and climate are different constructs, how can something be both weather and climate? The resolution is that one is just as dynamic as the other, and in both change is the norm. Mark Twain's "expectation" is also far more likely to pan out in some regions than others. At some location within the midlatitude storm track region, for example, where it is either very cloudy when a cyclonic system is passing overhead (let's call that 0.9) or mostly clear when an anticyclone is present (let's call that 0.1), there will be the same average as another place (e.g., in the deep tropics) where the cloudiness persistently hovers around 0.5. Perhaps we should just heed the words of Wally Broecker, of equal fame but in the field of geochemistry, that "The climate system is an angry beast, and we are poking at it with sticks."

5.4.1   The Global, Zonally Symmetric Perspective

The broadest features of the **general circulation** of the atmosphere include three cells that, together, will explain the global distribution of physical environments (and even biomes) such as those evident in the cloudiness maps presented in Figure 5.4, and derive their energy from the latitudinal structure of the net global energy imbalance as exposed in the previous section. As with the vertical layers of the atmosphere, we shall strive for understanding over memorization, so let's conduct another thought experiment.

Imagine a planet like "Earth" but without its namesake feature – land. It still has an atmosphere with four layers and an ocean everywhere beneath (Figure 5.5). It is also still a rotating sphere whose distance from the nearest radiant star (~150,000,000 km) is enormous compared to its distance around (~40,000 km). With such geometry, one horizontal strip along the sphere (the equator) will receive solar radiation at the most direct angle of incidence (perpendicular), so there will be more energy absorbed there, per square meter of surface area, than at the poles. Consequently, the equator will be hot and the poles cold. Since, gram for gram, hot air is more buoyant and tends to rise, there will be ascent at the equator and

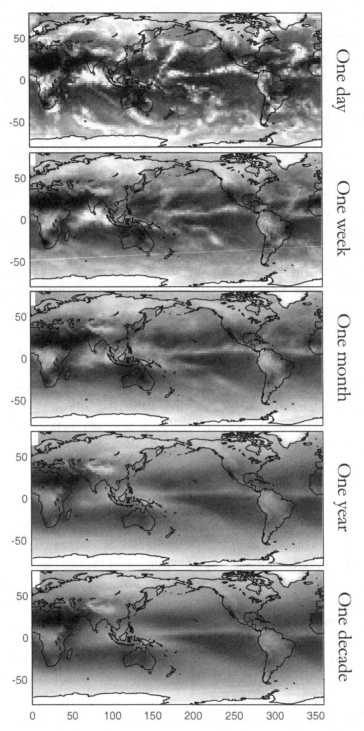

Figure 5.4 Maps of outgoing longwave radiation (OLR; closely related to cloud cover) averaged over one day (October 29, 2012), averaged over that week, that month, that year, and that decade. Lighter shades denote lower OLR (or more cloud cover); the same shading scale is applied to all five maps. OLR observations from the NOAA Interpolated OLR data set.

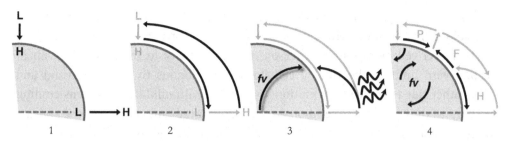

Figure 5.5 Illustration of the steps involved in the thought experiment building the three-cell model of the atmospheric general circulation and the prevailing surface wind patterns. In each step, black markings represent the new/relevant element(s).

descent over the poles (step 1). At the equator, as a result of the air evacuating the surface trying to push the tropopause upward with limited success (recall from Section 5.2 that air penetrating upward through the tropopause is difficult to sustain given the stable stratification), air pressure will be low at the surface and high near the tropical tropopause. With the opposite being the case at the poles (high surface pressure and low pressure aloft), a large-scale meridional pressure gradient $\partial p / \partial y$ will drive meridional flow that connects the single, hemispheric-scale over-turning cell (step 2). These are the physics that English polymath Edmund Halley (1656–1742) had in mind when he first attempted to explain the atmosphere's overturning in the tropics. It did not take us long to realize that this overturning cell originating in the tropics does not extend all the way to the poles. It has thus been named after George Hadley (1685–1768), largely in recognition of his making a more accurate connection between it and the trade winds at the surface.

Believe it or not, the mix of dynamics governing the poleward edge of the Hadley circulation is still hotly debated, especially as pertaining to the role of turbulent eddies, but there are a few basic tenets that surely contribute to its inability to reach the poles. First, the warm air that rises over the tropics and is thrust poleward radiates heat off to space, becomes cooler (and therefore denser), and eventually becomes negatively buoyant such that it sinks to the surface (step 3). Second, the planet is spinning, so the Coriolis force as described in the previous chapter will induce an eastward acceleration of the poleward-flowing air aloft, further making it difficult to reach the poles. The same will happen with the equatorward-flowing air at the surface originating at the poles; the strong Coriolis force at high latitudes will deflect that air westward long before reaching the equator. Finally, there is a bit of a logistical problem. Imagine all of the air encircling the planet at the equator, where the circumference is a full 40,000 km, trying to meet at the pole where the circumference approaches zero – there's not enough room. The gradual convergence of poleward-flowing air would increase the pressure aloft and eventually overcome the upward (hydrostatic) pressure gradient before reaching the pole. With our original overturning cell now clearly unable to reach the pole, yet the physics driving sinking air over the poles still present, a three-cell model emerges with **polar** (P) and **Hadley** (H) cells, and a so-called **Ferrel** (F) cell in between (step 4). The Ferrel cell owes its existence on one side simply to the descending air at the poleward limit of the Hadley cell (loosely analogous to a ball bearing) and on the other side to the vertical motions associated with midlatitude storms eroding the sharpest of the planet's meridional temperature gradients, such as those appearing as comma-shaped cloud formations in the top panel of Figure 5.4.

One can appreciate the importance of the atmospheric general circulation to the ocean simply by considering how it sets the broad-scale patterns of surface heat and freshwater flux terms as discussed in Chapters 2 and 3. Where abundant short-wave radiation reaches the surface of the ocean is strongly determined by the rising and sinking branches of the general circulation, particularly those of the Hadley cell. Likewise, the high water vapor content (a potent greenhouse gas) and cloudiness present in the deep tropics inhibits some of the ocean's emitted longwave radiation escaping rather than being absorbed and re-radiated back toward the ocean surface. Just beyond the tropics, relatively dry air with sufficient wind speed from the steady easterly trade winds fuels high rates of evaporation and latent heat flux – high sea surface salinity (SSS) is the result. To be sure, the atmospheric general circulation places a clear imprint on the thermodynamics of the ocean. The associated surface wind circulation also provides a vital *dynamical* forcing to the ocean by way of surface wind stress, the patterns of which will be described later in this chapter and the consequences of which will be examined in the next chapter.

On a planet such as the simplified one we conjured for this thought experiment, the three-cell model sketched in Figure 5.5 would be representative of any pole-to-pole slice of the atmosphere, at any longitude. The real world is, of course, not so simple since our planet *does* have continents. Nonetheless, its real-world counterpart retains much of the essential structure when heavily averaged (Figure 5.6). This is presented in the form of the *zonal mean meridional overturning streamfunction*, which is a calculation widely used in climate and atmospheric science to characterize the Hadley cell. Let's break that down. The phrase **zonal mean** simply indicates that the velocity data were first averaged across all longitudes, at each latitude (similar to the procedure employed to generate Figure 5.3d). "Meridional overturning" means that we are looking at a circulation in the y–z plane (i.e., flow in both the northward/southward and up/down directions). Finally, the Stoke's streamfunction, or just **streamfunction** for short, is a common mathematical construct used in fluid dynamics (including those of the ocean and atmosphere) to characterize a velocity field (and the associated transport of mass) in a two-dimensional plane (such as the coupling of meridional and vertical velocities, in this case). It can be quite useful because it is a scalar field, so it can distill two vectoral components into a single field, making many applications such as quantifying the strength or size of the Hadley cell easier. In this context, the streamfunction ψ and the meridional and vertical velocity components v and w are related, such that

$$v = -\frac{\partial \psi}{\partial z} \qquad (5.1a)$$

$$w = \frac{\partial \psi}{\partial y} \qquad (5.1b)$$

That is, flow will be directed northward where the streamfunction locally decreases with height z, and upward where streamfunction locally increases with latitude y.[2] The stronger those local gradients of ψ, the faster the flow (which equates here to more rapid transport of mass). The strong seasonality of the Hadley cell is evident by comparing the observed zonal mean meridional overturning streamfunctions for annual mean and boreal winter, with the upward branch of the Hadley cell, also known as the **ITCZ**, tending to reside in the summer hemisphere and the broad, descending branches in the winter hemisphere. That the ITCZ, including that of the annual mean climatology, is actually centered north of the equator is the first consequence of continents and their asymmetric distribution (more in the Northern Hemisphere) that we can spot. The Hadley cell is also subject to variations in the climate over time. For example, during an El Niño event, in which the SST of the equatorial Pacific warms by a couple of degrees Celsius, the Hadley cell is strengthened over the equator but contracts inward from its outer edges (Figure 5.6c).

It is not just a convenient coincidence that the seasonality of the Hadley cell is such that it is always transporting heat from the summer hemisphere to the winter hemisphere. Note that the polar and Hadley cells are **thermally direct**, such that they are driven by temperature gradients (hot air rising, cold surface air flowing equatorward, etc.) and thus they are major contributors to alleviating Earth's latitudinal energy imbalance, discussed in the previous section. The actual extent to which the atmosphere and ocean transport heat poleward to do so has been quantified by analysis of atmospheric and oceanic observations, as well as measurements of the net radiative balance at the top of the atmosphere by satellites (Figure 5.7). While the *direction* of the heat transports in both the atmosphere and ocean is poleward in both hemispheres, their magnitudes differ substantially. Outside of the tropics, the atmosphere contributes the vast majority of the poleward heat transport needed to balance the net radiative imbalance at the top of

---

[2] It is common practice in atmospheric science to use pressure p as a vertical coordinate rather than the Cartesian altitude z. In that case, vertical velocity is defined as $\omega = dp/dt$ rather than $w = dz/dt$, and since $\partial p/\partial z < 0$ in the atmosphere, the definition of streamfunction (Equation 5.1) emerges as $v = \partial \psi/\partial p$ and $\omega = -\partial \psi/\partial y$. Zonal mean meridional overturning streamfunction is computed by taking the vertical (pressure) integral of the zonal mean meridional velocity and applying an assumption that the atmosphere is nondivergent (i.e., the continuity equation is valid). Knowledge of the velocity field, which is much less well observed, is not required to calculate streamfunction – another advantage to the streamfunction approach of characterizing the overturning circulations in the atmosphere.

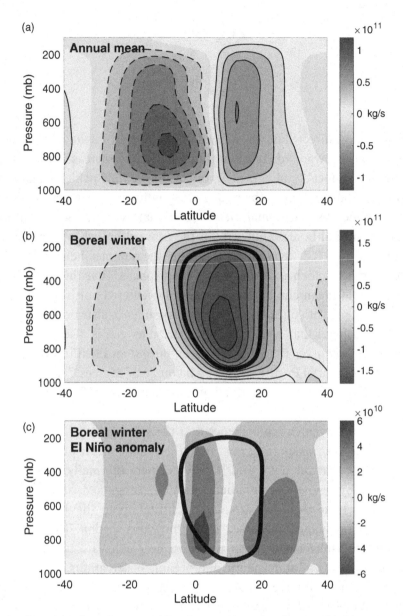

Figure 5.6 Climatological zonal mean meridional overturning streamfunction (kg/s) based on observational data (NOAA/NCEP-DOE Reanalysis 2 data set) for the annual mean (a), boreal winter (b) and the streamfunction anomaly during the 1997/1998 El Niño (c). Contour intervals in each panel are 2×10^{10} kg/s. For reference, the 10×10^{10} kg/s contour from the middle panel is repeated on the bottom panel as a thick black line. (A black and white version of this figure will appear in some formats. For the color version, please refer to the plate section.)

the atmosphere. Within the tropics, the partitioning is roughly equal between the atmosphere and ocean save for between the equator and ~10°N, where the ocean actually dominates.

For those keeping score, this method of partitioning the poleward heat transport between the atmosphere and ocean is, while logical, not entirely fair. For example,

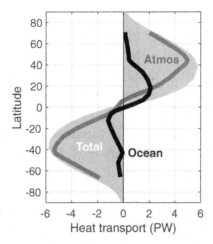

Figure 5.7 Estimated total poleward heat transport in petawatts (PW) as a function of latitude by the atmosphere (dark gray line) and ocean (black line), and the total required poleward heat transport based on the TOA energy imbalance (light gray shaded area). Poleward heat transport is defined as positive northward (so negative heat transport indicates southward). Plotted using data from Trenberth and Caron (2001).

the atmosphere gets credit for the substantial poleward heat transport achieved by tropical storms, but tropical storms are extracting a great deal of heat from the ocean throughout their life cycle, and the warmer and deeper the mixed layer of the ocean, the more intense hurricanes can become. In the end, the ocean and atmosphere both play for team climate, and heat is heat; it just has to get moved poleward one way or the other. Finally, it is interesting to notice how relatively symmetric the poleward heat transport is in the atmosphere (i.e., the Northern Hemisphere is a mirror image of the Southern Hemisphere). In contrast, the ocean's overall poleward heat transport is about twice as strong in the Northern Hemisphere than in the Southern Hemisphere, which foreshadows the importance of some regional aspects of the global ocean circulation that will be discussed in Chapter 8.

There is a deceptive neatness to the atmosphere both in terms of the zonal mean meridional overturning streamfunction and the latitudinal structure of its total poleward heat transport. There are in fact entire, and important, sectors of the troposphere where the circulation runs fully counter to that suggested by the previous three figures!

5.4.2   Zonal Asymmetry and Regional Circulations

Why does the atmospheric general circulation exhibit more variety than was portrayed in the previous section, even when heavily averaged? It all comes from the one aspect of Earth that we neglected in our thought experiment above: **continentality**. Continents, of course, provide separation between the ocean circulations within each basin, but they also provide a less visible separation between atmospheric circulations overlying each ocean basin. Without oceanic–continental boundaries, zonally uniform surface temperatures, a zonally uniform equatorial trough of low pressure, and a pair of zonally uniform subtropical high-pressure ridges would prevail. In the real world, continents *puncture* those otherwise continuous subtropical ridges with stationary, thermally driven low-pressure centers, leaving behind closed isobaric contours defining the stationary anticyclones over each subtropical ocean basin.

Examine the maps of some key climatic variables shown in Figure 5.8, beginning with the distribution of sea level pressure (P_{SL}). Notice how the subtropics in the Southern Hemisphere are *almost* blanketed by a continuous ridge of high pressure. Now look across the Northern Hemisphere; those discontinuities in the ridge are much more pronounced, and so too are the resulting subtropical anticyclones over the North Pacific and North Atlantic Oceans. This makes good sense because this is a picture of the average climatology for *boreal* summertime. As the major continental landmasses in the Northern Hemisphere, such as North America and northern Africa and the Middle East, heat up (high T_S) much faster than the stubborn ocean when given the same amount of incident shortwave radiation from the Sun, there is a greater tendency for tropospheric instability, rising air, and hence low-pressure centers to develop in place and break up the otherwise zonally symmetric high-pressure ridge. Such closed anticyclones over the oceans have implications for the fingerprint of the meridional overturning circulation of the atmosphere – the Hadley circulation. Due to the same principles of geostrophy that were discussed at the end of the previous chapter, there will be a tendency for *poleward* surface winds on the west sides of those subtropical anticyclones – that is, *opposite* to the direction of the canonical Hadley circulation! While we normally define positive v_{10} as northward, it is defined as equatorward in Figure 5.8 to highlight the regions where surface flow aligned with the Hadley circulation is actually occurring. To be clear, however, that we *have* a mean meridional overturning circulation (as clearly evident both schematically and in real-world data shown in the previous section), there is more than enough equatorward surface flow (and poleward flow aloft) to outweigh in the zonal mean such flows running counter to the Hadley circulation.

Another consequence of Earth's surface being filled with instances where oceans (high heat capacity) reside directly adjacent to land (low heat capacity) is that there are strong, seasonal heating gradients that drive the regional circulations we call **monsoons**. Like the Hadley and polar cells, most monsoons are also thermally direct – that is, they are effectively heat pumps driven by hot air rising. We can see evidence of some of the Northern Hemisphere monsoon systems in the surface and dynamical fields associated with annual mean climate (Figure 5.8). The most prominent monsoon on Earth is arguably the Indian Monsoon. Each monsoon system around the world has its own regionally specific details, but the Indian Monsoon is, like most, initiated as the landmass warms rapidly relative to the ocean as the seasons march into summer. This establishes instability in the lower troposphere, which enables air to start rising, cooling, condensing (i.e., raining),

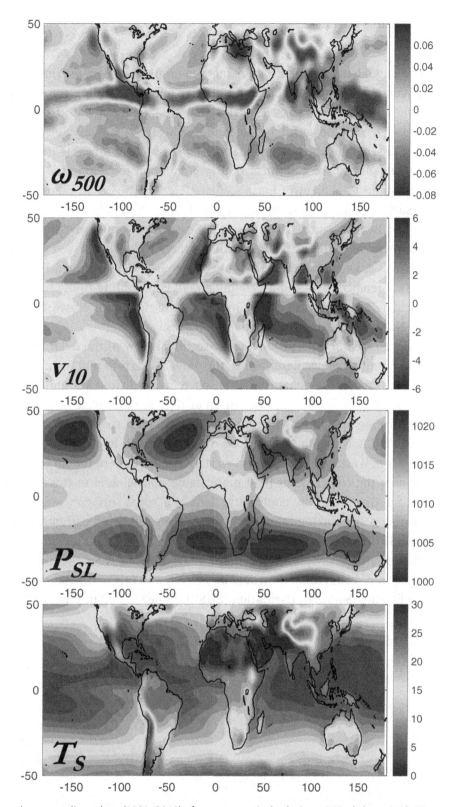

Figure 5.8 Boreal summer climatology (1981–2010) of pressure vertical velocity at 500 mb (ω_{500}; Pa/s), 10-m meridional wind (v_{10}; m/s), sea level pressure (P_{SL}; hPa), and surface temperature (T_s; equivalent to SST over ocean and land surface temperature [LST] over land; °C). Red indicates descending motion in ω_{500} and *equatorward* surface flow in v_{10}. Based on the NOAA/NCEP-DOE Reanalysis 2 data set. (A black and white version of this figure will appear in some formats. For the color version, please refer to the plate section.)

releasing latent heat, and then rising and raining some more. This process of **deep convection** facilitates building an upward plume extending throughout the depth of the troposphere, and even pushing the tropopause upward by a couple of hundred meters. So powerful are the land–sea heating gradients and deep convections over the Indian subcontinent that the Indian Monsoon is able to entirely override the tendency of the canonical Hadley circulation for a longitudinal stretch spanning some 15 percent of Earth's circumference! (Notice in Figure 5.8 the regions of surface wind running counter to the Hadley circulation in the Arabian Sea and Bay of Bengal, where air is being pulled toward southern Asia by the immense low-pressure center associated with the Indian Monsoon.) A very similar, yet less expansive, scenario can be seen over the southwestern United States, with warm surface temperatures and ascending air, all contributing to the relatively low continental P_{SL} that breaks up the high-pressure ridge into closed anticyclones over the North Pacific and North Atlantic Oceans.

If we take our pole-to-pole, cross-sectional view of the overturning circulation in the atmosphere as described in the previous section and turn it 90° so we have a slice through the equatorial atmosphere, we will find yet another set of coherent (as in usually present, not just an artifact of time-averaging), thermally direct overturning circulations (Figure 5.9). The broadest such zonal overturning circulation is known as the **Walker cell**, which is driven by deep convection fueled by the warm water over the tropical western Pacific/eastern Indian Ocean and sinking air over the cooler water (and more stable atmosphere) in the eastern equatorial Pacific. Given that, by continuity, there is a zonal surface wind implied between the rising and sinking branches of the Walker cell, wind can impact SST, and it is the SST distribution that drives the Walker cell, you might already suspect that the Walker cell is a key player in *coupled* interactions between the tropical atmosphere and ocean. Your suspicion would be correct, and this particular dance is known as the El Niño–Southern Oscillation (ENSO), to be studied in Chapter 7.

Figure 5.9 Schematic illustration of the time-averaged zonal overturning circulation in the tropics including the Walker cell over the Pacific Ocean. The red and blue shading represent warm and cold SSTs, respectively. Adapted from NOAA Climate.gov drawing by Fiona Martin. (A black and white version of this figure will appear in some formats. For the color version, please refer to the plate section.)

To the extent that we can consider today's geographic distribution of continents (and islands) "random," all of these regional features are just coincidences as well, but they definitely have a systematic impact on Earth's flora and fauna, including humans. Take the island nation of Indonesia and the Galápagos Archipelago (Ecuador), for example. These two landmasses are found at the same latitude but could not be more different in terms of climate and ecosystems. There are *penguins* on the Galápagos, for goodness sake! Interestingly enough, the arrangement of Indonesia as the hot rainforest and Galápagos as the cooler, desert island reverses entirely along with the Walker circulation during an El Niño event. Due to simultaneous changes in ocean circulation, the Galápagos penguin population drops off precipitously, too – but we'll get there soon enough in Chapter 7.

5.4.3 The Global Distribution of Surface Wind and Wind Stress

The prevailing surface winds on Earth emerge directly from the three-cell general circulation thought experiment conducted earlier in this section; the easterly trade winds in the tropics, the midlatitude westerlies, and the polar easterlies are all simply the surface manifestations of the three meridional overturning cells being deflected by the Coriolis force (as shown in step 4 of Figure 5.5). What the ocean feels directly, then, is a superposition of the zonal mean meridional overturning circulation discussed in the previous section, and the zonally asymmetric features described just above including the subtropical anticyclones, regional monsoons, and zonal overturning circulations such as the Walker cell (Figure 5.10). The equatorward surface flow of the Hadley cell can be seen, especially in the eastern

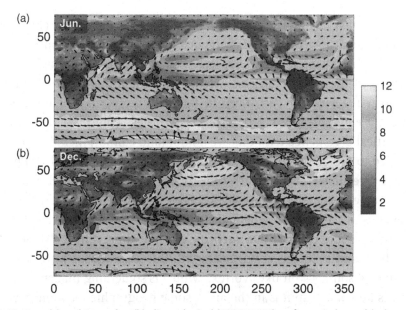

Figure 5.10 June (a) and December (b) climatological (1988–2017) surface wind speed (colors, m/s) and vectors. Based on the CCMP v.2 wind data set. (A black and white version of this figure will appear in some formats. For the color version, please refer to the plate section.)

halves of the three tropical ocean basins (except over the Indian Ocean during boreal summer when the monsoon dominates), the poleward flow in the western North Pacific and Atlantic during boreal summer can be seen, and the elongated zonal jets at ~50° latitude in either hemisphere – but especially during wintertime.

All of these features together determine the patterns of surface wind stress that act as a dynamic driver of the wind-driven component of the global ocean circulation. Interestingly, what we will quickly see by re-examining the ocean's momentum budget in terms of the wind's frictional effect on the oceans of a rotating planet like Earth is that the wind stress by itself is hardly the most useful quantity. Rather, there are a couple of key (partial) derivatives of the wind stress field that offer deeper insight and predictive power concerning ocean circulation, and those are its divergence and curl. So, before looking at those dynamics from the perspective of the ocean's momentum budget in the next chapter, let us reinterpret the map of surface wind (Figure 5.10) but in terms of wind stress and its derivatives.

As a reminder from the previous chapter, the wind *stress* is proportional to the wind speed squared (i.e., $\tau_x \propto u_a^2$), so it will be qualitatively the same but with an exaggerated dynamic range. First, we will define these wind stress derivatives, and then examine their geographic distribution. The divergence and curl of the wind stress vector [**X**] are defined as

$$\text{div}(\tau) = \nabla_h \cdot \tau = \frac{\partial \tau_x}{\partial x} + \frac{\partial \tau_y}{\partial y} \tag{5.2}$$

$$\text{curl}(\tau) = \nabla_h \times \tau = \frac{\partial \tau_y}{\partial x} - \frac{\partial \tau_x}{\partial y} \tag{5.3}$$

Even if you have a momentary lapse in recollection of vector calculus, they are not too difficult to remember. Divergence looks just like the continuity equation from the last chapter, both have $/\partial x$ then $/\partial y$, and everything else (which component is in the numerator, and whether there is an addition or subtraction in between) is reversed for curl. Keep in mind we're also not actually *violating* the continuity equation if $\nabla \cdot \tau \neq 0$, because that mess could be cleaned up by a nonzero vertical divergence term that we do not (and cannot) include here as wind stress applied to the ocean surface is only two dimensions (in x and y). Physically, **wind stress divergence** is just that – positive for a net horizontal export of air from some cubic meter situated at location (x,y), and negative for a net horizontal convergence of air therein. Mathematically, **wind stress curl** is what an atmospheric scientist would call **relative vorticity** when dealing with the standard wind components u_a and v_a in the vector operation. But it's a useful word to bring up anyway, since it sounds a bit like "vortex," which is something that spins. So, wind stress curl is how much spin is in the air at some geographic location (x,y).

Let's begin by taking in the real-world wind stress field and thinking through how these derivatives play out (Figure 5.11). The first thing that we might notice

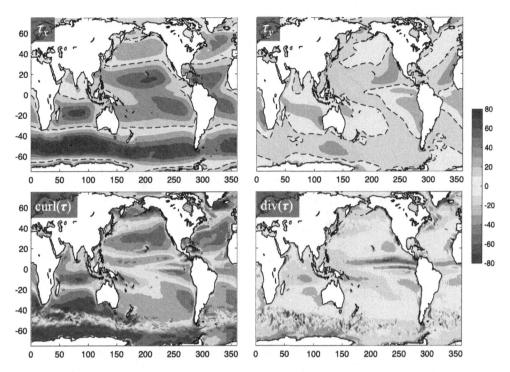

Figure 5.11 Annual mean (1988–2017) surface zonal wind stress (τ_x), meridional wind stress (τ_y), wind stress curl, and wind stress divergence. Wind stresses have units m^2/s^2 (pseudostress) and the curl and divergent fields are multiplied by 10^6 for scale purposes. Based on the CCMP v.2 wind data set. (A black and white version of this figure will appear in some formats. For the color version, please refer to the plate section.)

about the wind stress field is that its zonal component τ_x typically has larger values than the meridional wind stress τ_y, and that τ_x also has stronger variation in the y dimension (i.e., with latitude) than in the x dimension. In that light, one may just as well summarize the zonal wind stress map with a zonal mean profile (Figure 5.12) and note that there are two bumps of easterly wind stress in the tropics (one on either side of the equator) and two bumps of westerlies in the midlatitudes (about ± 30–$60°$ latitude) – all consistent with the setup of the atmospheric general circulation as described earlier in this chapter. Therefore, the wind stress curl is surely dominated by the $-\partial \tau_x/\partial y$ term and, by definition, has its most important features not on top of the strips of τ_x, but where they are transitioning most sharply between easterly and westerly (or between easterly and ~ 0, as is the case near the equator). So, on the map of wind stress curl, indeed we do find large curl signatures with opposite signs in both hemispheres centered around $30°N$ and $40°N$.[3]

---

[3] If you are looking at the map (or zonal mean profile) of zonal wind stress and are having difficulty associating these features with rotation, just imagine a pinwheel placed at $\sim 30°N$ in the Pacific (just north of Hawai'i). The midlatitude westerlies will blow the top side of the pinwheel to the east, and the tropical westerlies will blow the bottom side of the pinwheel toward the west, so

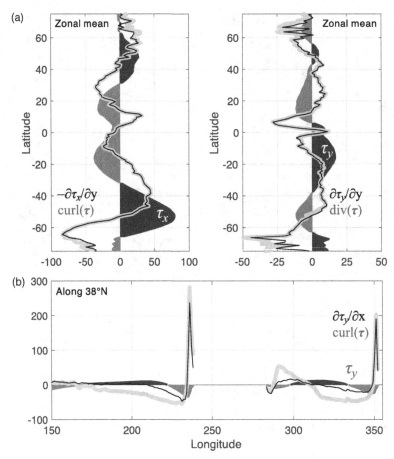

Figure 5.12 Annual mean (1988–2017) zonal mean profiles of wind stresses and derivatives thereof as indicated in the labeling (a). (b) Longitudinal profiles of annual mean meridional wind stress τ_y (red and blue profile), the zonal gradient of meridional wind stress $\partial \tau_y / \partial x$ (black line), and the full two-dimensional wind stress curl (gray line) along 38°N (roughly the latitude of San Francisco and Lisbon, Portugal). Wind stresses have units m^2/s^2 (pseudostress) and the curl and divergent fields are multiplied by 10^6 for scale purposes. Based on the CCMP v.2 wind data set. (A black and white version of this figure will appear in some formats. For the color version, please refer to the plate section.)

The divergence field is also one that faithfully reflects the structure of the atmospheric general circulation, including very strong negative wind stress divergence (i.e., convergence) along the equator where the ITCZ, or rising branch of the Hadley cell, ascends away from the surface. This is also quite clear in the zonal mean profile of meridional wind stress (Figure 5.12); the two equatorward surface flows meet a few degrees north of the equator, reflected in the very strong negative spike in wind stress divergence that is entirely contributed by its $\partial \tau_y / \partial y$ half. Although

the pinwheel will spin clockwise, or negative, which is anticyclonic in the Northern Hemisphere. Appreciable wind stress curl can, in fact, be generated by nothing but a nearby wind blowing in one direction, but of course it is more evident in two dimensions such as beneath a closed circulation like a tropical storm.

much weaker, one can also spot the broad regions across the subtropics where the Hadley (and Ferrel) cells descend, making those regions relatively sunny and dry (and contributing to the high SSS). Considering the zonal overturning circulations, one can also see the region in the western tropical Pacific where the tropical easterly trades slow down as they approach the rising branch of the Walker cell.

Despite the emphasis above on the meridional structure of the wind stress field and its leading role, particularly in setting the wind stress curl field, intriguing evidence of the ocean and atmosphere influencing one another can be seen in a sample *longitudinal* profile spanning the Pacific and Atlantic Oceans at roughly the latitude of San Francisco, California and Lisbon, Portugal (Figure 5.12b). In particular, near longitude 290°, one can see a substantial positive wind stress curl signature, which is associated with an influence from the Gulf Stream (more on this in the next section). The *very* large positive spike in wind stress curl in the eastern sides of both the Atlantic and Pacific, and composed entirely of the zonal gradient of meridional wind stress ($\partial \tau_y / \partial x$), is due to the abrupt weakening of the northerly wind at the respective continental margins. As we will see in the following chapter, these narrow but intense strips of positive wind stress that curl along the coastlines of Portugal and California explain much of what you'll see on these coastlines and their Southern Hemisphere counterparts: plenty of fish, and surfers in wetsuits even in the summer.

5.5   How the Atmosphere Responds to the Ocean

The global and regional features of the atmospheric general circulation presented in previous sections of this chapter are clearly influenced by the ocean. Our thought experiment began with rising air in the tropics driven by a warm surface, and ~78 percent of the circumference of Earth at the equator is ocean. Running perpendicular to the Hadley circulation, the Walker cell is driven by the zonal gradient of SST along the equator. Even the monsoon circulations, while thrust upward by the intense heating of the land surface relative to the ocean, are modulated by the land–sea temperature *difference* and fueled by the moisture that the inbound monsoonal winds are able to extract from the ocean through evaporative (latent) heat flux. Here we turn our attention to the ways in which the atmosphere responds to *variability* in the ocean, and for the most part at spatial scales smaller than the broad-brush features of the global atmospheric general circulation. It is a bit more complicated than "hot air rises," but it is not overwhelming either.

If we think about the actual means by which the ocean can, by itself, influence the atmosphere, it necessarily involves properties at the ocean–atmosphere interface (i.e., the surface of the ocean), and from there it usually boils down to SST. As we saw in Chapter 2, longwave radiation (Q_{LW}) and both of the surface turbulent heat fluxes Q_{LH} and Q_{SH} are strong functions of SST, so it is natural that SST is the key quantity here. Other avenues certainly exist, such as salinity's influence on surface fluxes (Q_{LW} and Q_{LH}) and swift ocean currents exerting a drag on the overlying atmosphere, but the former is much smaller than that of SST and the latter is small and patchy compared to the drag exerted by the atmosphere on the ocean (i.e., wind stress). Therefore, we can be more specific about the question at hand: How do SST anomalies influence the atmosphere? Despite its refinement, this question is surprisingly broad and challenging. The details remain vigorously debated among ocean/atmosphere scientists despite the substantial amount of research attention invested in it over the past three decades or so. Such intense interest in the subject is often rooted in recognition of its comprising one side of the coin of coupled ocean–atmosphere interactions.

There are five widely cited mechanisms by which variations in the ocean's SST field influence the atmosphere, as illustrated in Figure 5.13. One very influential paper published in 1980 by Adrien E. Gill (1937–1986) solved essentially the same equations of motion presented in the previous chapter (but applied to the atmosphere) with the addition of a localized heating term Q added to the continuity equation (Gill, 1980). His analytical solutions with Q centered on the equator rendered an astonishingly accurate depiction of the zonal overturning circulation in the tropics, including the Walker cell (label A in Figure 5.13). The premise of

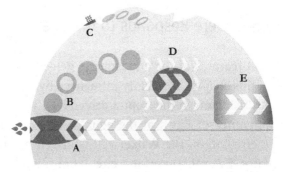

Figure 5.13 Schematic illustration of the different physical mechanisms by which the atmosphere responds to variability in the ocean. Warm SSTs near the equator force tropical deep convection, leading to a broad pattern of equatorial easterlies associated with an atmospheric Kelvin wave (A) and a stationary Rossby wave transmitting circulation anomalies into the midlatitudes (B). Reduced sea ice cover at high latitudes increases surface turbulent heat fluxes from the ocean to the atmosphere, which warms and moistens the boundary layer and similarly triggers a stationary Rossby wave with potential to impact storm tracks and the jet stream (C). Warm SSTs accelerate the prevailing surface wind by destabilizing the boundary layer and increasing the downward mixing of momentum to the surface (D). Cold SSTs have the opposite effect – to increase stratification, reduce vertical momentum mixing, and ultimately increase the vertical wind shear such that the winds are weaker at the surface. Horizontal *gradients* of SST induce horizontal gradients of temperature throughout a well-mixed boundary layer, which equate to hydrostatic pressure gradients that accelerate wind down the pressure gradient – or toward the warmer SST (E). (A black and white version of this figure will appear in some formats. For the color version, please refer to the plate section.)

the so-called Gill model is that his atmospheric heating term Q embodies warm SST accelerating the transfer of heat and moisture from the surface ocean to the lower atmosphere. This flux destabilizes the atmospheric boundary layer, driving air upward where it cools, condenses, and further releases (latent) heat. When this happens in a big, organized way, we call it deep tropical convection. The broad region of easterly surface winds converging toward the prescribed heating source was interpreted as atmospheric Kelvin waves (waves like this will be discussed in the next chapter). Not only might the Gill response (of the atmospheric circulation to SST) explain why we have a Walker circulation in the first place, but it also works well to explain what happens whenever those warm SSTs in the western equatorial Pacific get even warmer than normal, cool down, or move around. The tropical atmosphere responds just about every time, and the Gill model remains a useful theoretical framework by which we understand it.

Around the same time as Gill's groundbreaking paper, many were also thinking about how such convection and circulation anomalies in the tropical atmosphere might influence midlatitude weather and climate, and they were running more computer model experiments to investigate such poleward propagation of climate anomalies. The deep, upward motion in the atmosphere over warm SST anomalies that converges upon the tropopause must be balanced by divergence and subsidence somewhere else – and it doesn't just have to be over the eastern equatorial Pacific (i.e., the descending branch of the Walker cell). As early as 1966, the Norwegian meteorologist Jacob Bjerknes (1897–1975) was already testing the hypothesis (just using sparse observations!) that such signals make their way poleward via the Hadley circulation (Bjerknes, 1966). We now know this to be essentially true, but with the signal propagating poleward in an alternating sequence of high- and low-pressure systems, along with the ascent/descent and clockwise/counterclockwise circulation anomalies that go with them (label B in Figure 5.13) (Hoskins and Karoly, 1981; Trenberth *et al.*, 1998). These are also commonly interpreted as stationary atmospheric Rossby waves (more waves that will be discussed in the next chapter). When anomalous anticyclones and cyclones reach the midlatitudes, they superimpose onto the average features of the atmospheric general circulation and can have significant effects on phenomena of interest to midlatitude civilization like the position of storm tracks and the jet stream. Both the Gill model and the subsequent generation of stationary Rossby waves emanating into the midlatitudes depend critically on the initiation of deep convection, which is fueled by the latent heat released by the condensation of huge amounts of moist, ascending air. Therefore, these first two mechanisms are thought to be more applicable in tropical regions where the background SST is warm, due to the exponential dependence of water vapor-holding capacity on air temperature (i.e., the Clausius–Clapeyron relation – well known to students of meteorology).

Continuing our journey poleward, there are also unique impacts of ocean variability at high latitudes on the atmosphere. Sea ice acts as a cap to the surface fluxes introduced in Chapter 2, essentially shutting down the entire Q_{net} term since the bright white surface reflects most of the shortwave radiation, emits little long-wave radiation Q_{LW}, and prohibits entirely the surface turbulent heat fluxes Q_{SH} and Q_{LH}. When sea ice that is normally somewhere – like in the Arctic Ocean, around Antarctica, or other high-latitude seas such as the Sea of Okhotsk in the northwestern Pacific – is missing, heat and moisture can again be transferred from the ocean to the atmosphere (Honda *et al.*, 1996) (label C in Figure 5.13). Unlike the tropics, deep moist convection is improbable in the polar regions, but several studies have recently shown that the atmosphere does radiate the impacts of high-latitude surface warming far afield through a stationary Rossby wave such that the jet stream and midlatitude storm tracks can be altered by localized sea ice anomalies (Vihma, 2014; Barnes and Screen, 2015).

Even in the absence of deep convection like the driver of the classic Gill model, or atmospheric waves that transmit the signal of localized anomalies elsewhere, there are still two physical mechanisms by which the surface wind can be influenced by SST anomalies. One of them was discovered in a pair of studies out of Seattle, Washington, which were published simultaneously in 1989 (Hayes *et al.*, 1989; Wallace *et al.*, 1989). The idea is that SST influences surface air temperature through the surface heat fluxes, and the vertical profile of temperature in the atmosphere determines its stability, which determines how effectively the large momentum of the winds aloft will be mixed downward to the surface (label D in Figures 5.13 and 5.14). For example, a cold SST anomaly will reduce the rate at which temperature decreases with height, which increases the stability of the atmospheric boundary layer (overturning is less likely since the surface air is less buoyant), which reduces the turbulent mixing between the surface and winds aloft (which are faster due to their greater distance from surface friction). In other words, cold SST serves to *decouple* the surface from the free troposphere above the boundary layer, and vice versa for warm SST. By this rationale, we should expect to find wind speed positively correlated with SST (the warmer the SST, the faster the wind). This correlation has even been utilized as a diagnostic to determine whether ocean is driving the atmosphere, or the other way around. Imagine a case where you observed a negative correlation between SST and wind speed. That would be easier to explain with a mechanism like this: Faster wind speed increases surface turbulent fluxes, which cool the ocean mixed layer. Or, perhaps the fast wind is driving a circulation anomaly in the ocean that pulls cold water up from depth (a process called upwelling, which we will explore in the next chapter). Notice, however, that this mechanism has no impact on wind *direction*. In fact, this mechanism *only* applies to a situation in which the wind is already blowing; it does not involve one of the geophysical forces discussed in Chapter 4 that can set a fluid into motion. Despite the simplicity of this mechanism, and the fact that it was initially based on inferences drawn from a limited set of field measurements, it seems to be confirmed (if not

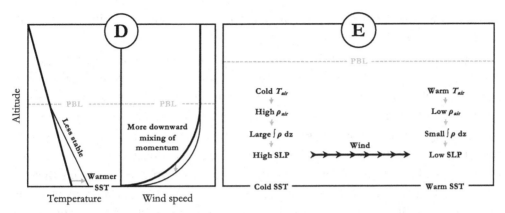

Figure 5.14 Schematic illustration of the physical arguments accompanying mechanisms D and E from Figure 5.13. See text for discussion.

refined) time and again – even using very different data sets such as global, high-resolution satellite observations. Notably, a set of papers out of the Oregon State University revealed in vivid detail the coherent patterns of accelerations and decelerations (and associated divergence patterns) across SST fronts such as those in the Gulf Stream and throughout the Southern Ocean (O'Neill *et al.*, 2003; Chelton *et al.*, 2004).

In stark contrast to the stability argument above, our fifth and final mechanism by which SSTs influence wind *does* invoke one of the fundamental geophysical forces: gravity via the pressure gradient force. A 1987 study out of the Massachusetts Institute of Technology, which remains perhaps one of the most widely discussed and debated among specialists, hypothesized that SST anomalies create horizontal (air) pressure gradients in between them, which naturally accelerate wind down the gradient (i.e., blowing from high to low pressure) (label E in Figures 5.13 and 5.14) (Lindzen and Nigam, 1987). Air temperature throughout the boundary layer is assumed to be in thermal equilibrium with SST, and the hydrostatic approximation is invoked to calculate pressure from temperature. That a pressure gradient would accelerate flow has not been questioned; the debate surrounding this mechanism is centered on some of the assumptions needed to arrive *at* a pressure field, given an SST field. For example, some have explored the questions of how well mixed the atmospheric boundary layer is, how deep (and rigid) the boundary layer is, and how much influence the free troposphere above has. In some cases, the breakdown of this model happens over *very* sharp SST gradients at relatively small spatial scales, such as meandering instability waves in the tropics, which seem to have provided inspiration for development of the stability argument (D) described immediately above. Others have improved the essential idea by including more physics that eliminates the need for such strong assumptions (Back and Bretherton, 2009). In the end, it cannot be denied that when a computer model simulation was built around this physical principle, it did a reasonable job reproducing the observed large-scale wind field, and so often (especially in the tropics) do we find even slight horizontal air pressure gradients driving substantial wind circulation.

To summarize, there is a wide variety of ways in which the atmospheric circulation is sensitive to thermodynamic forcing from the ocean. None of them are perfectly applicable in all situations. They are not necessarily mutually exclusive, either. If you look at a global map of surface wind on a given day, you are probably looking at a superposition of all five of these mechanisms (if not more). The scientific community has come to a conceptual understanding (with some lingering debate) of these mechanisms through an iterative process involving theory, modeling, field observation, and, as our capabilities in the Earth sciences have expanded, satellite remote sensing. The dissertation research of Larry O'Neill at Oregon State put conceptual hypotheses like D and E to rigorous tests using satellite observations and computer models, significantly expanding our understanding of the details of how patchy SST anomalies influence the atmosphere. Notice, too, that the physical processes discussed in this section are fraught with potential ocean–atmosphere coupled feedbacks – that is, the ocean impacts the wind, but as we are now about to address in the next chapter, the wind has a major influence on the ocean. Finally, in Chapter 7, we will put both of the one-way views (ocean affecting atmosphere, and atmosphere affecting ocean) together to understand and explore climate variability as a coupled *system* of interactions.

Further Reading

An excellent, wide-ranging overview of ocean–atmosphere interactions over cool SST regimes based primarily on satellite observations is given by Xie (2004).

Questions

1. What is the primary criterion on which the layers of the atmosphere are defined, and what is the magnitude of this characteristic *at* the tropopause?
2. Unlike the incoming solar radiation at the top of the atmosphere (TOA), which is strictly a function of latitude, the net radiation at TOA features some zonal variation. Using Figure 5.3, can you spot any regions where the net heat flux is substantially more or less than elsewhere along the same latitude? Can you pose a hypothesis to explain one of these deviations from the zonal mean?
3. Examine the global distribution of zonal and meridional wind stress, τ_x and τ_y, in Figure 5.11. Select one location on the map. (a) What is the approximate longitude and latitude, and how would you describe this location in the context of the global atmospheric general circulation? (b) Give the sign (positive,

negative, or zero) and a sense of scale (simply "big" or "small" is fine) of each of the following terms at this location: τ_x, τ_y, $\partial \tau_x/\partial x$, $\partial \tau_x/\partial y$, $\partial \tau_y/\partial x$, and $\partial \tau_y/\partial y$. (c) Determine the sign and general scale of the wind stress divergence and wind stress curl at this location. You can check your work against the bottom two panels of Figure 5.11. (d) Without reading ahead to Chapter 6, can you make a guess as to how the wind stress field might be influencing the ocean circulation at this location?

4. What is the sign of wind stress divergence associated with the Intertropical Convergence Zone (ITCZ)?

5. Examine the feature labeled D in Figure 5.13. Reproduce a rough sketch of the wind field (i.e., the symbols ">" of various sizes representing westerly winds of various magnitudes) on a piece of paper. Now, circle and label areas in the vicinity of this wind field where wind stress divergence (and convergence) will occur. Repeat but for areas of wind stress curl (cyclonic and anticyclonic).

6. The first two responses of the atmosphere to SST anomalies discussed in this chapter (i.e., the Gill model and poleward-propagating Rossby waves) are thought to be more efficient or likely to occur if the SST anomalies are super-imposed on a warmer background condition. Why is this so?

7. Compare and contrast the final two responses of the atmosphere to SST anomalies discussed in this chapter (i.e., the stability/momentum mixing mechanism and the horizontal SST/pressure gradient mechanism) in terms of the underlying assumptions and the general contexts in which they apply.

8. Share your thoughts on the dichotomy between weather and climate. Are they clearly separable constructs? Is it more of a continuum? Is climate best described as the statistical mean of weather data, or is the relationship more complex? In what way(s) might weather and climate be *interactive*? Provide a specific example.

9. Select one of the *Dive into the Data* boxes featured in this chapter. Read the sample journal article that uses the associated data set, and describe the general role of that data set in the study. Provide a little context, both technical and scientific. What did the authors actually "do" with the data set? What specific scientific insight was enabled by incorporating this data set into the study?

10. Utilizing *Dive into the Data* Boxes 5.3 and 5.4, produce global maps of the time-mean sea level pressure (SLP) and outgoing longwave radiation (OLR). (a) Qualitatively describe the Walker circulation in terms of the spatial distribution of SLP and OLR in the tropics. (b) Offer a physical explanation for why, especially in the tropics, OLR is observed to be low wherever SLP is also low. (c) Select two locations somewhere in the world ocean – one in the tropics and one in the higher latitudes. How do OLR and SLP covary over time at these two locations? How does this temporal relationship between SLP and OLR differ between tropical and higher-latitude regions? Can you offer an explanation?

11. Utilizing *Dive into the Data* Box 4.1, calculate and produce a global map of the time-averaged meridional wind stress, τ_y. Refer to Equation 4.10 for a method to calculate wind stress τ_y from wind speed v_a. Assume the surface air density $\rho_a = 1.2 \, \text{kg/m}^3$ and the drag coefficient $C_d = 0.0013$, and be careful to retain the sign of τ_y! (a) Select a tropical region and identify the latitude of the ITCZ at that longitude using a criterion developed based on τ or one of its derivatives. (b) How does the ITCZ latitude vary as a function of season? (c) Are there any regions (or longitude ranges) where such an objective criterion would struggle severely to identify the latitude of the ITCZ? How so?

DIVE INTO THE DATA BOX 5.1

Name	Western Tropical Pacific Merged Ocean–Atmosphere Temperature Profiles
Synopsis	One full decade of daily atmospheric soundings (temperature) from five Micronesian islands (Palau, Yap, Truk, Ponape, and Majuro) situated within the western Pacific warm pool region combined with daily vertical profiles from nearby TAO moorings into a single profile for each day spanning 2002–2011. Raw soundings acquired from the University of Wyoming archive (http://weather.uwyo.edu/upperair/sounding.html). All daily soundings (one from each island) and profiles (one from each mooring) are averaged into a single profile representing the vertical thermal structure of the coupled ocean–atmosphere system in the western tropical Pacific region.
Science	Combined vertical profiles of temperature in the ocean (including the thermocline) and atmosphere (from the surface through the troposphere and into the stratosphere) can be used to investigate the vertical structure of tropical temperature anomalies (such as those associated with El Niño), the exchange of heat between the ocean and atmosphere, troposphere–stratosphere coupling and the vertical propagation of waves.
Figure	5.1
Version	N/A
Variable	Temperature
Platform	Radiosonde and mooring
Spatial	750 m depth to 30 km altitude, variable spacing in ocean (25–250 m), 10 m spacing in atmosphere, western tropical Pacific region
Temporal	2002–2011, daily averages
Source	See article
Format	MATLAB (.mat)

DIVE INTO THE DATA BOX 5.1   (cont.)

Resource   moat.m; moat.mat (sample data and code provided on publisher website here: www.cambridge.org/karnauskas)

Journal Reference for Data
Zhang, L., Karnauskas, K. B., Weiss, J. B., and Polvani, L. M. Observational evidence of the downstream impact on tropical rainfall from stratospheric Kelvin waves. *Clim. Dyn.* **50**, 3775–3782 (2018). DOI: 10.1007/s00382-017-3844-1.

Sample Journal Article Using Data
Zhang, L., Karnauskas, K. B., Weiss, J. B., and Polvani, L. M. Observational evidence of the downstream impact on tropical rainfall from stratospheric Kelvin waves. Clim. Dyn. **50**, 3775–3782 (2018). DOI: 10.1007/s00382-017-3844-1.

DIVE INTO THE DATA BOX 5.2

Name	CERES Top of Atmosphere (TOA) Energy Budget
Synopsis	Satellite observations of the TOA energy fluxes, including incoming solar radiation, reflected (outgoing) solar radiation, and emitted (outgoing) longwave radiation. Net TOA flux is then calculated as the difference between incoming and outgoing fluxes.
Science	The TOA energy budget reflects overall changes in the climate from shorter-term fluctuations (e.g., ENSO) to longer-term trends (greenhouse gas forcing). The circulation of the ocean and atmosphere exert a strong control on the TOA energy budget, since they both transport excess heat from the tropics to the poles. The structure of the TOA energy budget can also be used to understand these transports, and to evaluate the realism of global climate models.
Figure	5.3
Variable	Energy flux
Platform	Satellite
Version	4, EBAF (Energy Balanced and Filled)
Spatial	Global (gridded), 1° resolution
Temporal	Climatology over July 2005 to June 2015, monthly averages
Source	https://ceres.larc.nasa.gov/order_data.php
Format	NetCDF (.nc)
Resource	ceres.m; ceres.mat (sample data and code provided on publisher website here: www.cambridge.org/karnauskas)

DIVE INTO THE DATA BOX 5.2 (cont.)

Journal Reference for Data

Loeb, N. G., Doelling, D., Wang, H., *et al.* Clouds and the Earth's Radiant Energy System (CERES) Energy Balanced and Filled (EBAF) Top–of–Atmosphere (TOA) Edition-4.0 data product. *J. Climate* **31**, 895–918 (2018). DOI: 10.1175/JCLI-D-17-0208.1.

Sample Journal Article Using Data

Dessler, A. E. and Forster, P. M. An estimate of equilibrium climate sensitivity from interannual variability. *J. Geophys. Res. Atmos.* **123**, 8634–8645 (2018). DOI: 10.1029/2018JD028481.

DIVE INTO THE DATA BOX 5.3

Name	NOAA Interpolated Outgoing Longwave Radiation (OLR)
Synopsis	Outgoing longwave radiation measured directly by a series of NOAA satellites.
Science	Outgoing longwave radiation is a major component of Earth's TOA energy budget. This global data set, now spanning nearly 45 years, enables studies of climate variability across a wide range of temporal and spatial scales. It is also a useful perspective on rainfall, since towering tropical clouds that produce rain are cold and thus emit less OLR.
Figure	5.4
Version	N/A
Variable	Energy flux
Platform	Satellite
Spatial	Global (gridded), 2.5° resolution
Temporal	1975 to present, daily and monthly averages
Source	www.esrl.noaa.gov/psd/data/gridded/data.interp_OLR.html
Format	NetCDF (.nc)
Resource	olr.m; olr.mat (sample data and code provided on publisher website here: www.cambridge.org/karnauskas)

Journal Reference for Data

Liebmann, B. and Smith, C. A. Description of a complete (interpolated) outgoing longwave radiation dataset. *Bull. Am. Meteorol. Soc.* **77**, 1275–1277 (1996).

DIVE INTO THE DATA BOX 5.3 (cont.)

Sample Journal Article Using Data

Karnauskas, K. B. and Li, L. Predicting Atlantic seasonal hurricane activity using outgoing longwave radiation over Africa. *Geophys. Res. Lett.* **43**, 7152–7159 (2016). DOI: 10.1002/2016GL069792.

DIVE INTO THE DATA BOX 5.4

Name	NOAA/NCEP-DOE Reanalysis 2
Synopsis	An atmospheric reanalysis is an atmospheric general circulation model that is anchored by real observations. The method of anchoring, or "constraining," a model is known as data assimilation (i.e., assimilating real data into a model). Atmospheric (and oceanic) reanalyses are important tools/data sets for climate research because we usually do not have observations everywhere and from every time in the past – especially if one is interested in weather or climate conditions several decades ago. Reanalyses essentially fill the gaps, but with a model's physical equations rather than simply interpolating. These are among the most widely used data sets in the atmospheric/climate sciences community.
Science	The NOAA/NCEP-DOE Reanalysis 2 can be used to study almost any large-scale feature of the global atmosphere over the past 40 years from the surface to the tropopause. The reanalysis data set includes dozens of variables at all levels of the atmosphere, such as temperature, humidity, wind, cloudiness, energy fluxes, and more. Precipitation from atmospheric reanalyses should always be used with extreme caution.
Figures	5.6, 5.8, 6.7, 7.7, 7.9
Version	N/A
Variable	Multiple (e.g., 300 mb geopotential height)
Platform	Data assimilation
Spatial	Global (gridded), 2.5° resolution
Temporal	1979 to present, six-hourly, daily and monthly averages
Source	www.esrl.noaa.gov/psd/data/gridded/data.ncep.reanalysis2.html
Format	NetCDF (.nc)
Resource	ncep2.m; ncep.mat (sample data and code provided on publisher website here: www.cambridge.org/karnauskas)

DIVE INTO THE DATA BOX 5.4 (cont.)

Journal Reference for Data

Kanamitsu, M., Ebisuzaki, W., Woollen, J., *et al.* NCEP-DOE AMIP-II Reanalysis (R-2). *Bull. Amer. Meteor. Soc.* **83**, 1631–1644 (2002). DOI: 10.1175/BAMS-83-11-1631.

Sample Journal Article Using Data

Mitas, C. M. and Clement, A. Has the Hadley cell been strengthening in recent decades? *Geophys. Res. Lett.* **32**, L03809 (2005). DOI: 10.1029/2004GL021765.

6 Response to Wind Forcing

6.1 Overview of Wind Forcing

The previous chapter concluded with some of the ways in which the atmospheric circulation is influenced by the ocean. Since the ocean tends to move much more slowly than the atmosphere, the ocean's "drag" on the speedy and chaotic atmosphere is only significant in limited contexts. Rather, the focus was on the atmosphere's *thermodynamic* adjustment to SST variability. As we now turn our attention to the other side of the coin of ocean–atmosphere interaction, we have the occasion to examine a *dynamic* response to forcing. Not only is wind stress a very significant driver of the overall global ocean circulation, its dynamics and physical manifestations in the real ocean are also so readily distinguishable from all other factors that we can generally view phenomena that we observe as either part of the wind- or buoyancy-driven ocean circulation – the latter of which will come later.

This was the stuff of legends: Ekman at the turn of the century, Sverdrup, Stommel, and Munk in the roaring (nineteen) forties of physical oceanography. The turn of the millennium saw further advancement in our understanding of the wind-driven ocean circulation – and by a more diverse crowd of oceanographers like Sarah Gille, who put those early theories to the test using modern observations like satellites and on more extreme frontiers like the Roaring Forties of the Southern Ocean (Gille *et al.*, 2001). Perhaps the best part is that even the 1905 paper by V. Walfrid Ekman (1874–1954), published in *Arkiv För Matematik, Astronomi och Fysik* (Ekman, 1905), is actually readable; all you need is some familiarity with the equations of motion as presented in Chapter 4! Those were indeed his starting point – he didn't invent them, but his clever manipulations of them led to astonishing insight, which set us on course for the comprehensive theory and intuitive understanding of the wind-driven ocean circulation that we still enjoy testing today.

This chapter's presentation of the ocean's response to wind forcing is divided into two perspectives: steady state and transient. After grasping the dynamics of steady-state wind forcing, we will link them back to certain concepts like geostrophy and the heat budget to prove that the great gyres swirling about broad sea level hills in the subtropical and subpolar oceans are actually wind-driven! A set of waves that are of particular importance in coupled climate variability – ones capable of carrying the message of local climate anomalies clear across an ocean basin – will also arise naturally from blending these concepts.

6.2   The Ocean's Response to Steady Wind Forcing

6.2.1   Ekman Dynamics

Just as Ekman set out to in 1905, let's find out how a steady wind propels water in the upper ocean – that is, from the surface down to some level where the wind's viscous influence becomes immaterial. A natural place to start is the equations of motion and some simplifications consistent with the aforementioned goal:

$$\frac{\partial u}{\partial t} = -\frac{1}{\rho}\frac{\partial p}{\partial x} + fv + A_h \nabla_h^2 u + A_z \frac{\partial^2 u}{\partial z^2} - \mathbf{V}\cdot\nabla u \tag{6.1a}$$

$$\frac{\partial v}{\partial t} = -\frac{1}{\rho}\frac{\partial p}{\partial y} - fu + A_h \nabla_h^2 v + A_z \frac{\partial^2 v}{\partial z^2} - \mathbf{V}\cdot\nabla v \tag{6.1b}$$

Now let's rephrase the above horizontal momentum budget, borrowed from Chapter 4 (Equation 4.19), neglecting the time derivative (steady state), horizontal pressure gradient force (flat sea surface), horizontal diffusion, and advection (homogeneous flow). It's rarely defensible to neglect the Coriolis force here on Earth, so that term shall remain. Moreover, let's invoke the equivalence between vertical diffusion and wind stress (Section 4.2.3), yielding simply

$$-fv = \frac{1}{\rho}\frac{\partial \tau_x}{\partial z} \tag{6.2a}$$

$$fu = \frac{1}{\rho}\frac{\partial \tau_y}{\partial z} \tag{6.2b}$$

The above equations, implying a balance between the Coriolis force and the frictional influence of the surface wind stress, can be called the **Ekman balance.** They're simple, yes, but not quite ready to be applied to something tangible.

Just like we did when making use of the geostrophic balance in Chapter 4, we can capitalize on our good fortune that, even after eliminating the time derivative of the velocities, there are still velocities surviving even in these bare-bones equations because they were coupled with the Coriolis frequency f. So, let's solve for them (simply by dividing through by f), and then integrate both sides from the surface down to $z = z_{Ek}$, where z_{Ek} is the depth at which the influence of the wind is negligible. Doing so yields

$$U_{Ek} = \frac{\tau_y}{\rho f} \tag{6.3a}$$

$$V_{Ek} = -\frac{\tau_x}{\rho f} \tag{6.3b}$$

whereby not only did we eliminate the partial (vertical) derivative on the wind stress components, but we have new terms U_{Ek} and V_{Ek} that represent the vertically integrated horizontal velocity, or **volume transport**. Volume transports have units m³/s and are usually expressed by oceanographers in *millions* of cubic meters (of seawater) passing through a horizontal plane per second; they are named "Sverdrups" or Sv for short – we'll meet Harald Sverdrup later in this chapter. In this case, since U_{Ek} and V_{Ek} fell out of the Ekman balance, we call them **Ekman transports** and assign them a subscript "*Ek*" so we don't forget the assumptions under which they are applicable.

What does Equation 6.3 tell us? For a positive f (i.e., in the Northern Hemisphere), if there is a positive meridional (northward) wind stress, we predict a positive zonal (eastward) Ekman transport, and if there is a positive zonal (eastward) wind stress, we predict a negative meridional (southward) Ekman transport. In other words, the good old "90° to the right/left in the Northern/Southern Hemisphere" mantra applies here, too. This is another powerful tool; Equation 6.3 allows one to make a prediction of the bulk movement of seawater in the upper layer simply by knowing the speed and direction of the wind (or better, the wind stress).

In practical applications of Ekman dynamics and, specifically, Ekman transports to diagnosing the large-scale ocean circulation and even the inner workings of *coupled* climate variability, we rarely need to worry about the details of the currents *within* the so-called **Ekman layer**. However, it is worth taking a step back and considering how the viscous influence of the wind evolves as it penetrates downward and, for that matter, just where this Ekman layer ends. The whole derivation, as first performed by Ekman in 1905, was actually inspired by an observation made several years earlier by polar explorer Fridtjof Nansen (1861–1930) that icebergs tended to flow at an angle of 20–40° relative to the wind direction. Assuming Nansen's observations are robust, then surely Ekman's theory wouldn't suggest that the whole Ekman layer is moving uniformly at 90° to the right of the wind – and it doesn't.

Recognizing the "obvious" form of Equation 6.1 (with vertical diffusion $A_z \, \partial^2 u/\partial z^2$ and $A_z \, \partial^2 v/\partial z^2$) as a partial differential equation (PDE) with a general solution that predicts $u = v$ at $z = 0$, he deduced that the angle of the surface current should be 45° to the right of the wind (in the Northern Hemisphere). This mathematical prediction of Ekman's wasn't far off from Nansen's observation; recall that some portion of an iceberg is above the waterline and consequently subject to some amount of direct wind forcing (like a sail) rather than having its momentum borrowed entirely from the sea. Moreover, the general solution to that PDE predicts that the magnitudes of u and v both depend on z such that the total current vector *rotates* continuously with depth like a spiral, and the scalar speed $|\mathbf{V}|$ *decays exponentially* with depth away from the surface. This phenomenon is now famously known as the **Ekman spiral** (Figure 6.1), and the depth at which the velocity has rotated a full 180° from the surface is typically considered the **Ekman layer depth** (z_{Ek}).

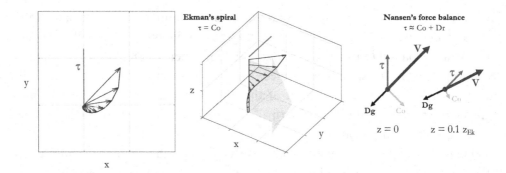

Figure 6.1 Birds-eye and 3D views of the Ekman spiral beneath a steady northward wind stress (medium gray line), and a reinterpretation of it (just the top two vectors shown in the Ekman spiral) based on Nansen's force balance (Dg = drag, Co = Coriolis force). In the 3D view of the Ekman spiral, the transparent arrow represents the net (Ekman) transport.

For the record, Nansen had actually described (in admittedly qualitative terms) a balance of forces leading to a spiral-like structure of the currents beneath the wind, which remains useful for solidifying our conceptual understanding of the essential physics (even if not quite the mathematics) behind Ekman's spiral. A reinterpretation of the top two layers of the Ekman spiral (at the surface and 10 percent of the way down to z_{Ek}) through Nansen's line of thinking is also shown in Figure 6.1. At the surface, the water is being propelled by the wind stress, deflected to the right by the Coriolis force, and held back by "drag" within the water. Those three forces balance (i.e., their vector sum equals zero) when the current is at some angle to the right of the wind (in the Northern Hemisphere). According to Ekman's solution, that angle is exactly 45°. For the next layer beneath the surface, the surface current now plays the role that the wind stress did previously; the surface current imparts its stress on the water below, which is similarly deflected to the right by the Coriolis force and decelerated by drag. Repeating this process over and over, the current vector continues to rotate *and* slow with depth, eventually (at z_{Ek}) directed very slowly in the opposite direction as the surface current. Ekman's mathematical manipulations of the horizontal momentum budget, which led off this section, showed that the *net* sum (or vertical integral) of all of those velocities – the Ekman transport – works out to be 90° relative to the surface wind.

If you ask a meteorologist where to find the most interesting atmospheric phenomena, they are likely to say something about vertical motion. Storm chasers aim for regions with rising air, while sinking air goes with dry, pleasant weather. Oceanographers from across the various disciplines including marine biology, and those interested more broadly in the interactions between the ocean and atmosphere, are likewise interested in vertical motions within the ocean. With such sharp vertical gradients of temperature and nutrients near the surface, even a relatively

Figure 1.3 [Year: 1962, distance: 266 km] Photo of Earth taken by astronaut John Glenn on February 20, 1962 aboard the Mercury spacecraft during his flight as the first American to orbit the Earth. Credit: NASA/ John Glenn.

Figure 1.4 [Year: 1972, distance: 29,000 km] Photo of Earth, dubbed "Blue Marble," taken by the crew of Apollo 17 on December 7, 1972 while en route to the Moon. Photo credit: NASA/Apollo 17 crew.

Figure 1.5 [Year: 1990, distance: 5,954,572,800 km] Photo of Earth, dubbed "Pale Blue Dot," taken by the Voyager 1 space probe upon leaving the Solar System on February 14, 1990. Credit: NASA/Voyager 1.

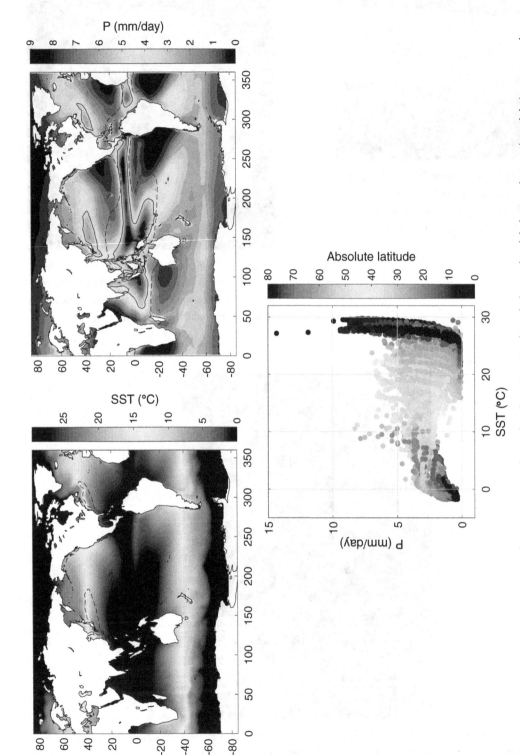

Figure 1.8 Maps of time-averaged sea surface temperature (SST; °C) and precipitation (mm/day) over the global ocean (top row), and their corresponding value of the absolute value of latitude (bottom panel). On the SST map, the 27 °C and 5 mm/day contours are shown in solid and dashed, respectively, and vice versa on the precipitation map. SST and precipitation observations from the NOAA OIv2 and GPCP data sets, respectively; both averaged from 1982 through 2018.

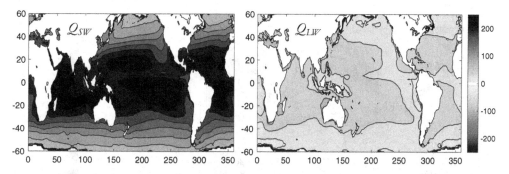

Figure 2.2 Global maps of net surface shortwave Q_{SW} and longwave Q_{LW} radiation (W/m²) averaged from 1984 to 2009 from the WHOI OAFlux data set. Positive (negative) values represent a net flux of radiant energy into (out of) the ocean. Solid lines are contoured every 25 W/m².

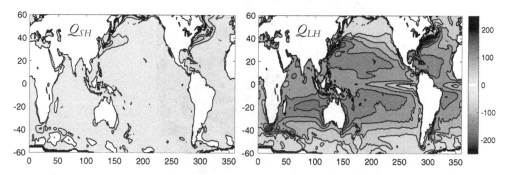

Figure 2.5 Global maps of surface sensible Q_{SH} and latent Q_{LH} turbulent heat flux (W/m²) averaged from 1984 to 2009 from the WHOI OAFlux data set. Positive (negative) values represent a net heat flux into (out of) the ocean. Solid lines are contoured every 25 W/m².

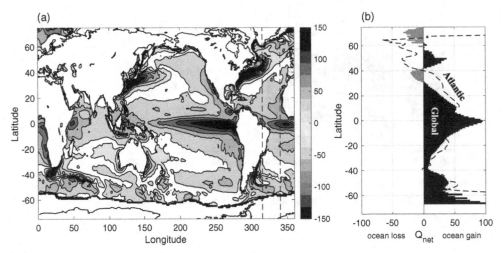

Figure 2.6 (a) Global map of net heat flux into the ocean Q_{net} (W/m²) averaged from 1984 to 2009 using the WHOI OAFlux data set. (b) Profiles of Q_{net} zonally averaged across all longitudes (colored bars) and across only the central Atlantic Ocean (dashed) as marked by dashed lines on the map.

Figure 3.1 Photo of an Argo float being deployed in 2014 by then-graduate student Tyler Hennon on cruise US GO–SHIP P16S in the Southern Ocean under the direction of Chief Scientist Lynne Talley. Associated with the Southern Ocean Carbon and Climate Observations and Modeling (SOCCOM) project, the float pictured is instrumented for biogeochemical measurements such as dissolved oxygen, nitrate, and pH, in addition to the traditional physical variables. Photo credit: Isa Rosso, Scripps Institution of Oceanography.

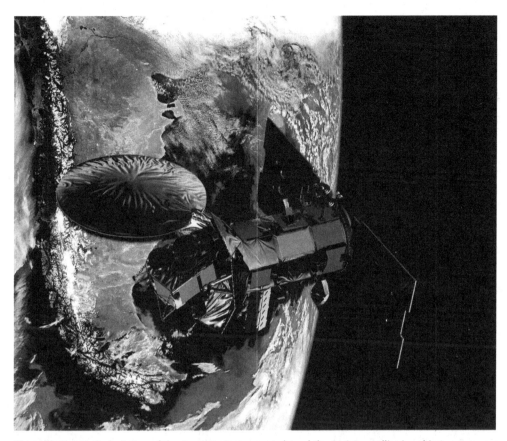

Figure 3.2 Artist's depiction of the Aquarius instrument aboard the SAC-D satellite in orbit. Image credit: NASA.

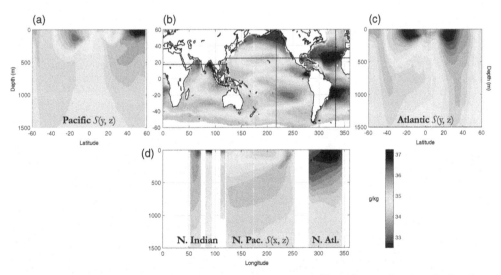

Figure 3.3 Map of SSS (g/kg) from Aquarius satellite observations (b), with cross-sections of subsurface salinity from Argo profiles from 60°S to 60°N through the Pacific (a) and Atlantic (c) Oceans, around the world at ~20°N (d). White areas in the bottom cross-section are landmasses. All data averaged over the same time period (2012–2014). The Roemmich–Gilson Argo Climatology was used.

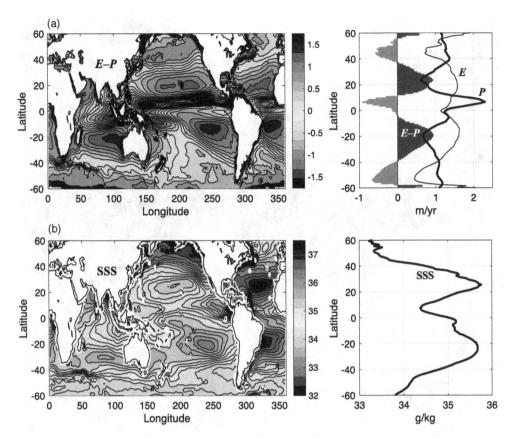

Figure 3.5 (a) Freshwater flux $E - P$ (meters per year) averaged from 1984 to 2009 using the WHOI OAFlux and GPCP data sets and profiles of E, P, and $E - P$ zonally averaged across all longitudes. (b) As in (a), but for SSS (g/kg) averaged from 2012 to 2014 using the Aquarius data set.

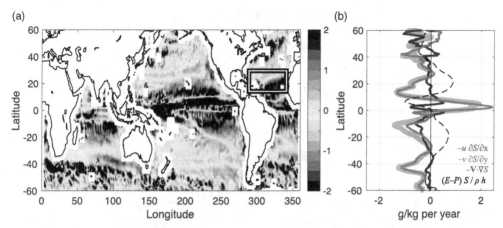

Figure 3.7 (a) Horizontal salinity advection (sum of zonal and meridional advection terms $-u\,\partial S/\partial x$ and $-v\,\partial S/\partial y$) (g/kg per year) using surface velocities from the SODA data set averaged from 1984 to 2008 and SSS from the Aquarius data set averaged from 2012 to 2014. The inset box marks the domain of the Figure 3.8. (b) Profiles of ocean mixed layer salinity budget terms zonally averaged across all longitudes, with spatially varying mixed layer depth in the freshwater flux term based on a fixed density threshold criterion of 0.03 kg/m³.

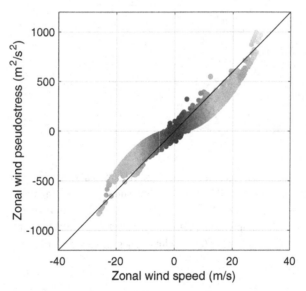

Figure 4.3 Illustration of the relationship between wind speed and wind stress (using pseudostress [m²/s²] from the Cross-Calibrated Multi-Platform [CCMP] level 3.5a data set). Each point on the graph represents one ¼° grid cell over the world ocean for one of the 12 months of the year 2000. There are a total of ~6.5 million points.

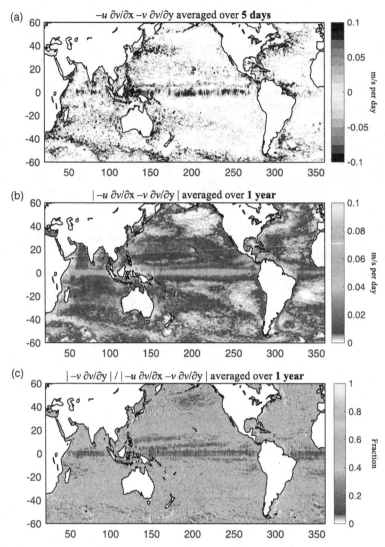

Figure 4.4 Horizontal advection of meridional velocity ($-u\,\partial v/\partial x -v\,\partial v/\partial y$) during the last pentad of 2018 from 1/3° OSCAR surface currents (a), the absolute value of $-u\,\partial v/\partial x -v\,\partial v/\partial y$ averaged over 2018 (b), and the ratio of meridional advection of meridional velocity ($-v\,\partial v/\partial y$) to the total horizontal advection of meridional velocity (c) (absolute values, averaged over 2018).

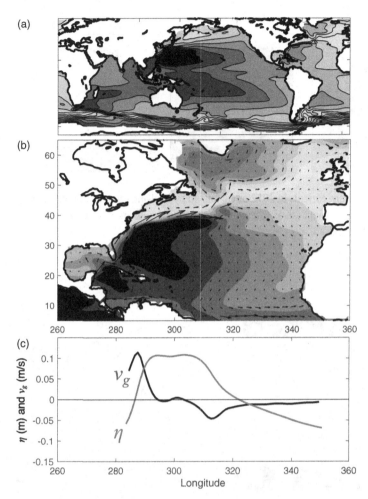

Figure 4.6 Sea surface height above geoid η averaged from 1993 to 2010 from the 1° AVISO satellite altimeter product with contour interval 0.2 m and values spanning roughly −1.5 to 1.5 m (a). Closer view of η and calculated geostrophic velocity vectors in the North Atlantic region with η contour interval 0.1 m and values spanning roughly −0.6 to 0.9 m (b). Longitudinal profiles of η and v_g along 36.5°N latitude (c). To align the two profiles, η was plotted as $0.3\eta - 0.1$.

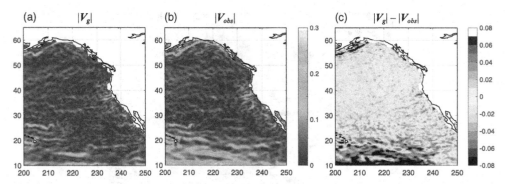

Figure 4.7 Near-surface currents in 2012 as calculated by applying the geostrophic velocity equations to observed sea surface height (a), as estimated by the OSCAR satellite product (b) and their difference (c). The units in all panels are m/s.

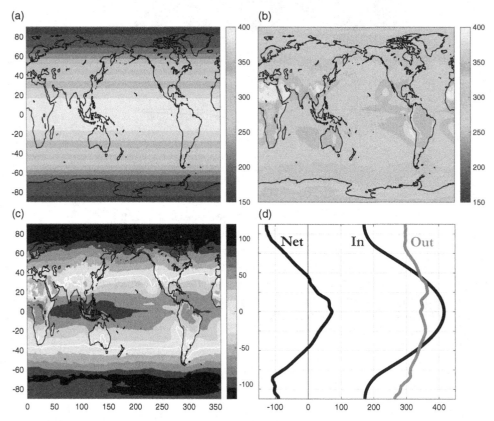

Figure 5.3 Long-term mean energy fluxes (W/m²) at the top of the atmosphere (TOA). Incoming shortwave radiation (a), total outgoing radiation (reflected solar plus longwave) (b), net radiation (incoming minus outgoing) (c), and zonal means of each quantity (d). The zero net radiation contour is shown in white in (c). Energy fluxes averaged from July 2005 through June 2015 from the NASA Clouds and Earth's Radiation Energy System (CERES) data set.

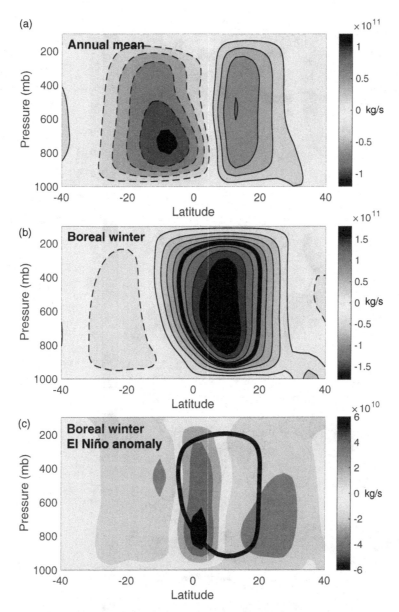

Figure 5.6 Climatological zonal mean meridional overturning streamfunction (kg/s) based on observational data (NOAA/NCEP-DOE Reanalysis 2 data set) for the annual mean (a), boreal winter (b) and the streamfunction anomaly during the 1997/1998 El Niño (c). Contour intervals in each panel are 2×10^{10} kg/s. For reference, the 10×10^{10} kg/s contour from the middle panel is repeated on the bottom panel as a thick black line.

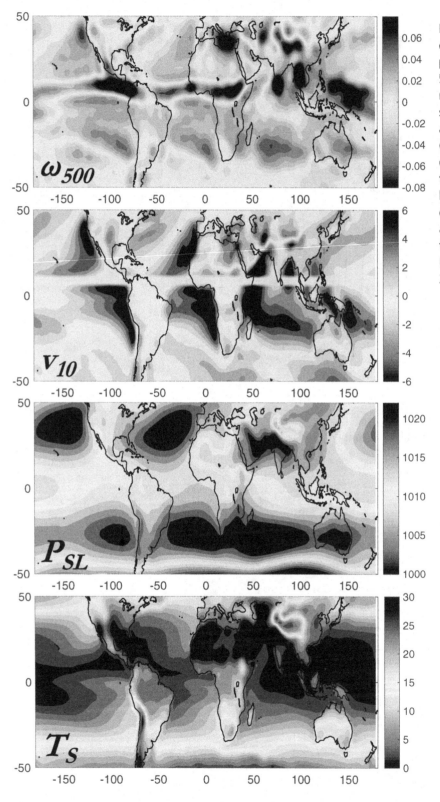

Figure 5.8 Boreal summer climatology (1981–2010) of pressure vertical velocity at 500 mb (ω_{500}; Pa/s), 10-m meridional wind (v_{10}; m/s), sea level pressure (P_{SL}; hPa), and surface temperature (T_s; equivalent to SST over ocean and land surface temperature [LST] over land; °C). Red indicates descending motion in ω_{500} and *equatorward* surface flow in v_{10}. Based on the NOAA/NCEP-DOE Reanalysis 2 data set.

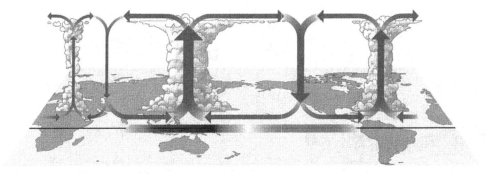

Figure 5.9 Schematic illustration of the time-averaged zonal overturning circulation in the tropics including the Walker cell over the Pacific Ocean. The red and blue shading represent warm and cold SSTs, respectively. Adapted from NOAA Climate.gov drawing by Fiona Martin.

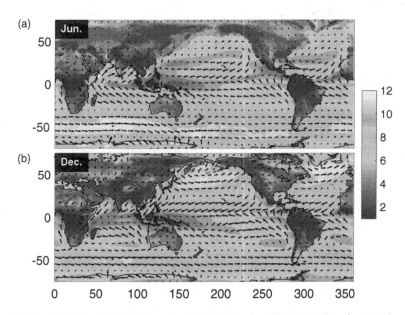

Figure 5.10 June (a) and December (b) climatological (1988–2017) surface wind speed (colors, m/s) and vectors. Based on the CCMP v.2 wind data set.

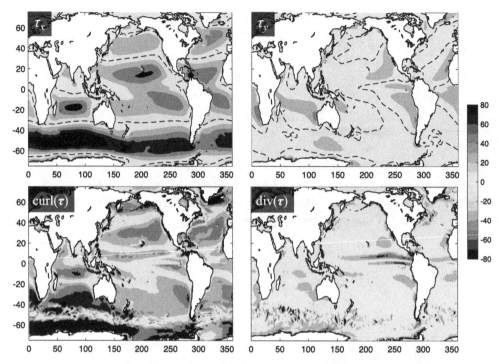

Figure 5.11 Annual mean (1988–2017) surface zonal wind stress (τ_x), meridional wind stress (τ_y), wind stress curl, and wind stress divergence. Wind stresses have units m^2/s^2 (pseudostress) and the curl and divergent fields are multiplied by 10^6 for scale purposes. Based on the CCMP v.2 wind data set.

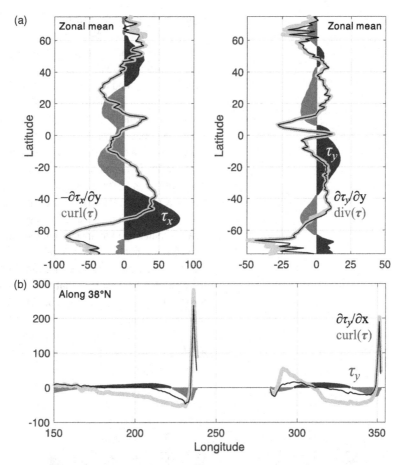

Figure 5.12 Annual mean (1988–2017) zonal mean profiles of wind stresses and derivatives thereof as indicated in the labeling (a). (b) Longitudinal profiles of annual mean meridional wind stress τ_y (red and blue profile), the zonal gradient of meridional wind stress $\partial \tau_y/\partial x$ (black line), and the full two-dimensional wind stress curl (gray line) along 38°N (roughly the latitude of San Francisco, USA and Lisbon, Portugal). Wind stresses have units m^2/s^2 (pseudostress) and the curl and divergent fields are multiplied by 10^6 for scale purposes. Based on the CCMP v.2 wind data set.

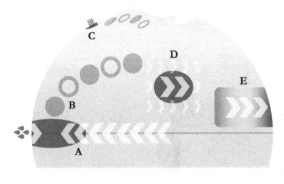

Figure 5.13 Schematic illustration of the different physical mechanisms by which the atmosphere responds to variability in the ocean. Warm SSTs near the equator force tropical deep convection, leading to a broad pattern of equatorial easterlies associated with an atmospheric Kelvin wave (A) and a stationary Rossby wave transmitting circulation anomalies into the midlatitudes (B). Reduced sea ice cover at high latitudes increases surface turbulent heat fluxes from the ocean to the atmosphere, which warms and moistens the boundary layer and similarly triggers a stationary Rossby wave with potential to impact storm tracks and the jet stream (C). Warm SSTs accelerate the prevailing surface wind by destabilizing the boundary layer and increasing the downward mixing of momentum to the surface (D). Cold SSTs have the opposite effect – to increase stratification, reduce vertical momentum mixing, and ultimately increase the vertical wind shear such that the winds are weaker at the surface. Horizontal *gradients* of SST induce horizontal gradients of temperature throughout a well-mixed boundary layer, which equate to hydrostatic pressure gradients that accelerate wind down the pressure gradient – or toward the warmer SST (E).

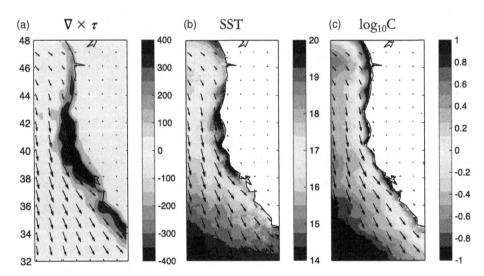

Figure 6.3 Wind stress curl (a), sea surface temperature (b), and the logarithm (base 10) of chlorophyll concentration (c) in the northeastern Pacific Ocean (along the west coast of the USA). In each panel, the wind stress is also shown. Data are from the boreal summer climatology (June–August) from the following satellite products: CCMP v.2 winds and MODIS Aqua.

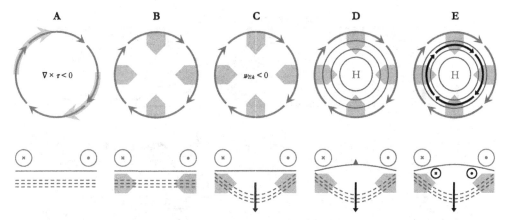

Figure 6.5 Schematic illustration of the mechanisms involved in the wind forcing of the subtropical ocean gyres, valid for the Northern Hemisphere. Wind in blue, sea surface height in solid red, subsurface isotherms (thermocline) in dashed red, Ekman transports in gray, Ekman pumping velocity in black, and geostrophic current in black. The thick transparent arrows (shown only in step A) represent the surface branches of the Hadley and Ferrel cells.

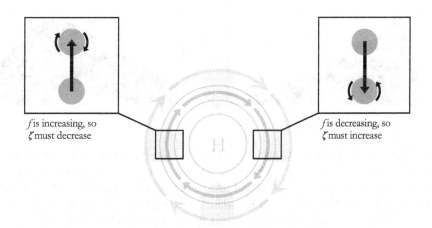

Figure 6.6 Schematic illustration of the conservation of potential vorticity amid meridional flows on either side of a subtropical ocean gyre, valid for the Northern Hemisphere. The faded element in the center of the illustration is step E from Figure 6.5.

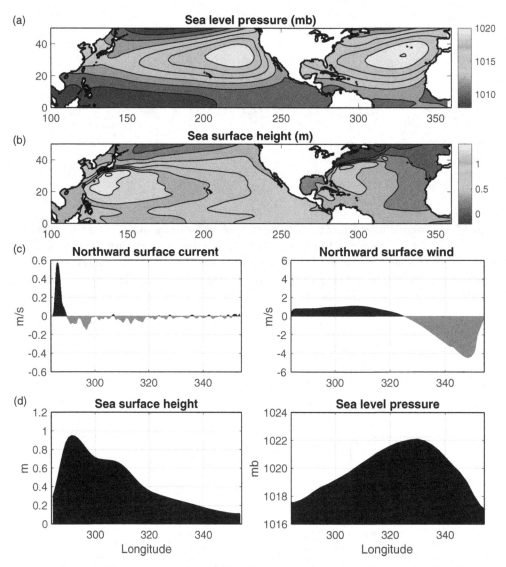

Figure 6.7 Maps of long-term annual mean (a) sea level pressure (SLP) and (b) sea surface height (SSH) in the North Pacific and Atlantic. (c) Profiles of meridional surface velocity in the ocean (left) and atmosphere (right) along ~35°N in the North Atlantic. (d) As in (c), but profiles of SSH (left) and SLP (right). SLP data from the NOAA/NCEP-DOE Reanalysis 2 data set (2.5°, 1988–2017), SSH data from the AVISO product (1°, 1993–2010), surface currents from the OSCAR product (1/3°, 1993–2010) and surface winds from CCMP v.2 product (1/4°, 1988–2017).

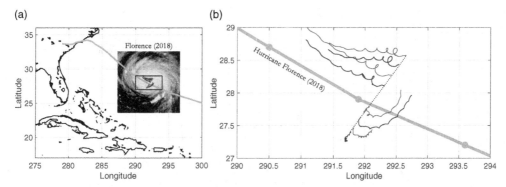

Figure 6.8 Inertial oscillations generated by Hurricane Florence in 2019. The overview map (a) shows the path and scale of Florence relative to the deployed floats. The inset map (b) shows the detailed trajectories of the floats during and after the passage of Florence. Image of Florence taken by NOAA/GOES-16 satellite and processed by the Cooperative Institute for Research in the Atmosphere (CIRA). Data from the Air Launched Autonomous Micro Observer (ALAMO) float program courtesy of Elizabeth Sanabia, US Naval Academy and Steven Jayne, Woods Hole Oceanographic Institution.

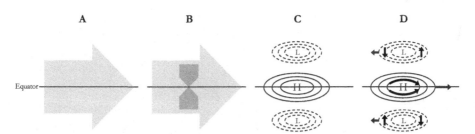

Figure 6.9 Schematic illustration of the production of Kelvin and Rossby waves by a temporary westerly wind stress centered on the equator.

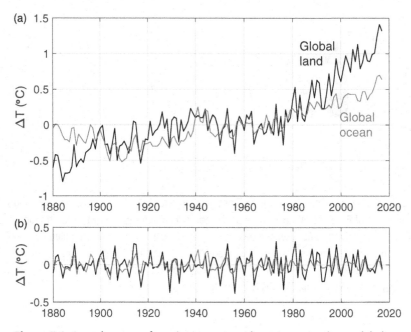

Figure 7.1 Annual mean surface air temperature changes averaged over global land (red) and ocean (blue) areas from 1880 to 2017. (a) The full record; (b) the record with the five-year smoothed records removed, to highlight only the year-to-year variations without the long-term trend. Data from the NASA GISS Surface Temperature Analysis (GISTEMP). All temperatures relative to 1951–1980 base period.

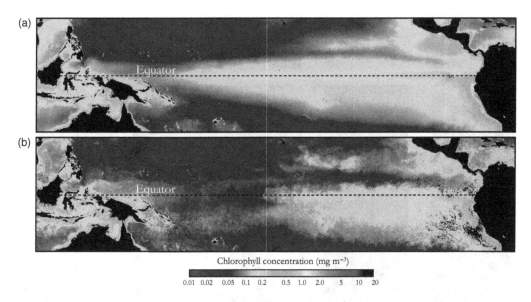

Figure 7.3 Surface chlorophyll concentration in the equatorial Pacific region during boreal autumn for average conditions (2002–2017 (a)) and during an El Niño event (2015 (b)), from the NASA MODIS instrument aboard the Terra satellite. Non-land areas colored black are due to cloud contamination of the satellite data.

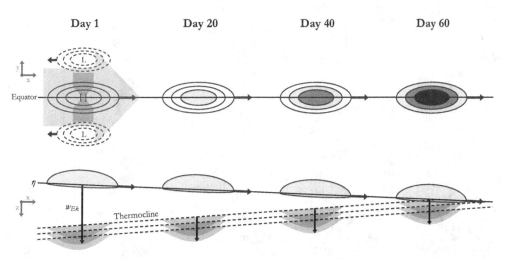

Figure 7.5 Schematic illustration of an equatorially trapped Kelvin wave propagating eastward across the Pacific, warming SST where the mean thermocline is shallow enough.

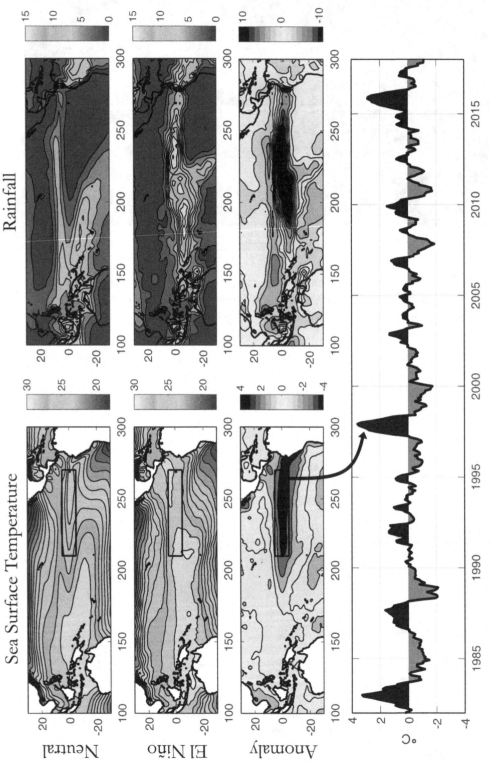

Figure 7.6 Maps of (left) SST (°C) and (right) rainfall (mm/day) for boreal winter neutral conditions (1982–2018 average), El Niño (December 1997–February 1998), and the anomaly (difference between El Niño and neutral). The bottom panel shows a time series of SST anomaly in the NIÑO3 region (i.e., averaged over the eastern equatorial Pacific between 150°W–90°W and 5°S–5°N; box drawn on the SST maps). SST and rainfall observations from the NOAA OIv2 and GPCP products, respectively.

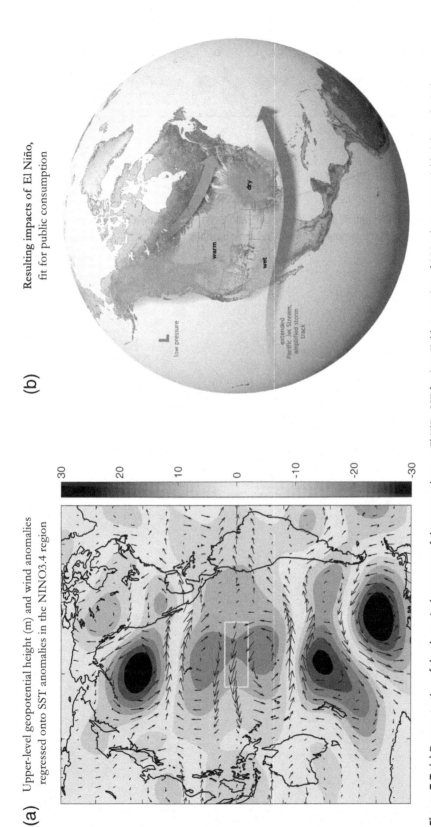

(a) Upper-level geopotential height (m) and wind anomalies regressed onto SST anomalies in the NINO3.4 region

(b) Resulting impacts of El Niño, fit for public consumption

Figure 7.7 (a) Demonstration of the dynamical response of the troposphere to El Niño SST forcing. Field regression of 300 mb geopotential height and wind anomalies onto the time series of SST anomalies in the NIÑO3.4 region (170°W–120°W and 5°S–5°N; box drawn on map). Units for 300 mb geopotential height are meters per standard deviation of variability in the NIÑO3.4 index; positive values indicate anomalous highs or anticyclones and negative values indicate anomalous lows or cyclones. (c) Schematic meant to communicate the likely impacts of El Niño to the general public on the boreal wintertime jet stream and North American surface climate (NOAA Climate.gov).

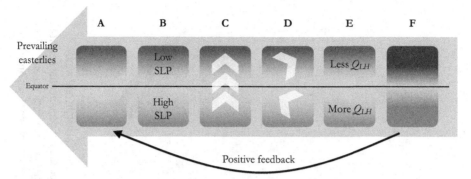

Figure 7.8 Schematic illustration of the Wind–Evaporation–SST (WES) feedback, a positive feedback in the tropics that is capable of amplifying an anomalous meridional SST gradient.

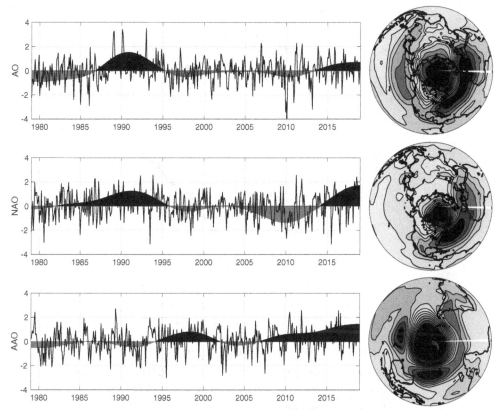

Figure 7.9 Arctic Oscillation (AO, aka Northern Annular Mode [NAM]), North Atlantic Oscillation (NAO), and Antarctic Oscillation (AAO, aka Southern Annular Mode [SAM]). Monthly time series (left column) from 1979 to 2018 with a 10-year smoothing and sea level pressure (SLP) anomaly correlations (right column). The SLP anomaly correlation maps have contour interval 0.1 and the colors saturate at correlation coefficients of ±0.66. Climate indices obtained from the NOAA Climate Prediction Center (CPC), and SLP data from the NOAA/NCEP-DOE Reanalysis 2 data set.

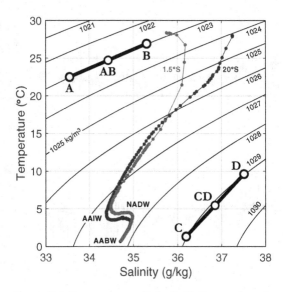

Figure 8.1 Temperature–salinity (T–S) diagram, based
on the equation of state, giving the seawater density
ρ (kg/m³) for a wide range of temperatures (0–30°C)
and salinities (33–38 g/kg) for pressure at mean sea
level (1013.25 hPa). Heavy black features illustrate
cabbeling, or the process of two water masses with the
same density mixing to create a water mass of greater
density. Two hydrographic profiles are shown from the
South Atlantic Ocean (WOCE Line A16C, R/V Melville,
13 March to 19 April 1989, Chief Scientist Lynne Talley):
one at 25°W, 20°S (blue) and one at 25°W, 1.5°S
(red). Temperature and salinity values plotted roughly
every 10 m depth. Water masses labeled include North
Atlantic Deep Water (NADW), Antarctic Intermediate
Water (AAIW), and Antarctic Bottom Water (AABW).

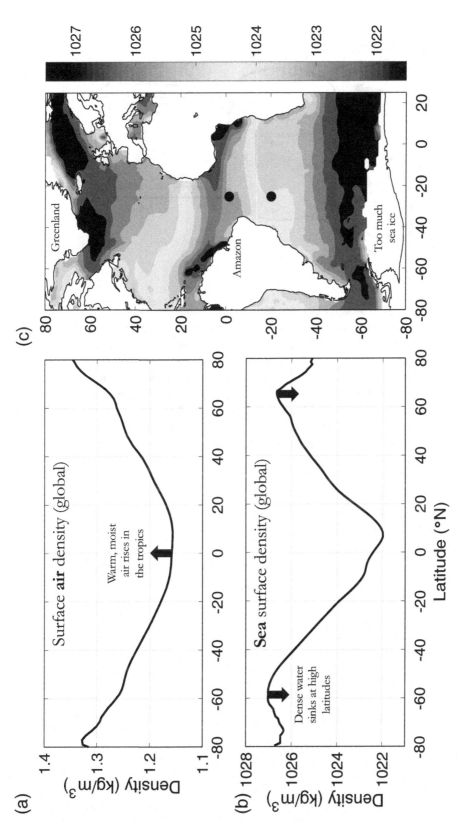

Figure 8.3 Zonal mean profiles of surface air density (a) and sea surface density (b) averaged over the period 2012–2014. (c) A map of sea surface density (kg/m³) in the Atlantic sector for the same period. Sea surface density calculated using the equation of state based on observed sea surface temperature (NOAA OIv2) and sea surface salinity (Aquarius). The locations of the two hydrographic profiles shown in Figure 8.1 are marked as black circles on the map.

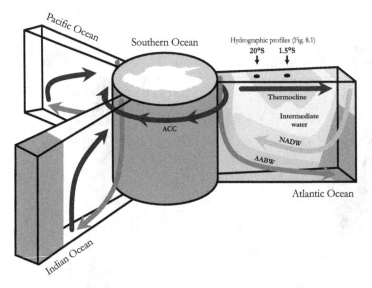

Figure 8.4 Schematic illustration of the global thermohaline circulation with an Antarctic-centered view of interbasin flows, inspired by the diagrams of Arnold Gordon (1991) and William J. Schmitz, Jr. (1996). Water mass details shown in the Atlantic sector include the thermocline, intermediate water, North Atlantic Deep Water (NADW), and Antarctic Bottom Water (AAB). The Antarctic Circumpolar Current (ACC) flows clockwise (from west to east) around Antarctica, connecting the three major ocean basins. The locations of the two hydrographic profiles shown in Figure 8.1 are labeled on top of the Atlantic sector of the diagram.

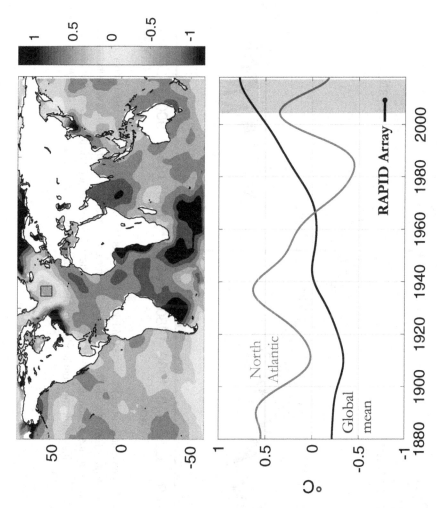

Figure 8.9 Map of linear trend in SST over the period 1880–2018 (°C per century) (a). Time series of SST anomaly in the North Atlantic Ocean (averaged over the small box indicated on the trend map) (blue line) and global mean surface temperature anomaly (red line). Both time series are smoothed with a 30-year low-pass filter. SST data from the NOAA Extended Reconstructed SST version 5 (ERSTv5) data set and global mean surface temperature from the NASA GISS Surface Temperature Analysis (GISTEMP) data set. All temperature anomalies relative to the 1951–1980 base period.

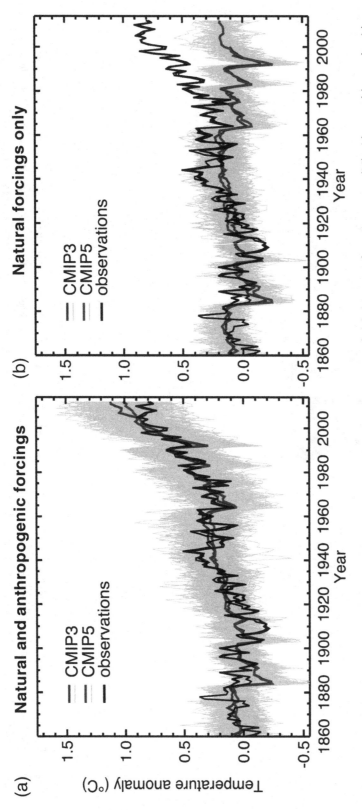

Figure 9.4 Historical simulations by ensembles of GCMs of the CMIP3 and CMIP5 generation of global mean surface temperature (°C) with natural (e.g., volcanic) and anthropogenic (e.g., greenhouse gases) forcings (a). The plot in (b) is the same except that only natural forcings were included in the model simulations, thus *attributing* the observed temperature change to anthropogenic forcings. Three different observational records are shown in black lines. Individual model simulations are shown in thin light blue and yellow lines, while the multi-model average is indicated by thick blue and red lines for CMIP3 and CMIP5, respectively. Adapted from IPCC AR5 WG1, Figure 10.1 (IPCC, 2013).

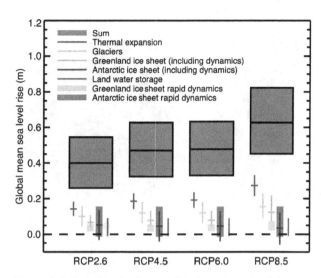

Figure 9.6 Future projections of global mean sea level rise (m) at the end of the twenty-first century (2081–2100) relative to the recent past (1986–2005) under four Representative Concentration Pathways (RCPs) by GCM simulations and process models. Adapted from IPCC AR5 WG1, figure 13.10 (IPCC, 2013).

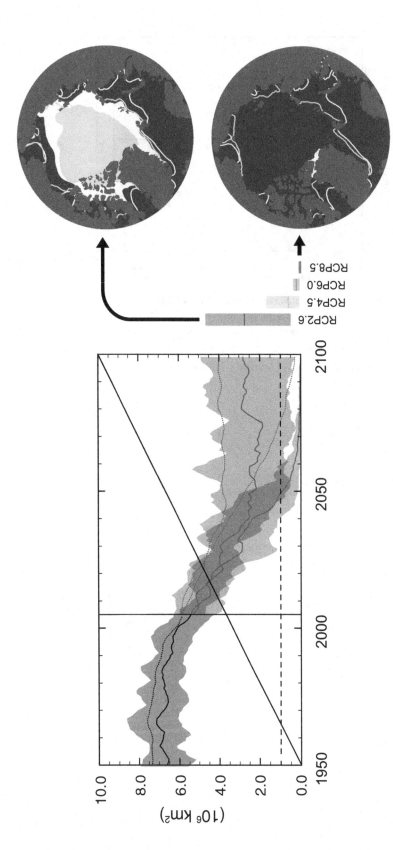

Figure 9.7 Historical simulations and future projections of sea ice extent in the Arctic during September (the seasonal minimum). Shown on the left are the projections under two Representative Concentration Pathways: RCP2.6 and RCP8.5, including the inter-model spread. On the right are the inter-model spreads averaged over 2081–2100 for all four RCPs and Northern Hemisphere maps, indicating the CMIP5 multi-model average over 1986–2005 (white lines) and the CMIP5 multi-model average projections for 2081–2100 (white area). The light blue line and area indicate the results for a subset of CMIP5 models that most closely reproduce the climatological pattern and recent observed trend (1979–2012) of Arctic sea ice extent. Adapted from IPCC AR5 WG1, figure SPM.7 and SPM.8 (IPCC, 2013).

modest ascending current, for example, can dramatically cool the sea surface temperature (SST) (to which the atmosphere may respond) and inject enough nutrients into the sunlit euphotic zone to fuel photosynthesis, supporting a persistent ecosystem. It's unclear who first did so, but it is a natural extension of the above Ekman dynamics, simply by invoking the continuity equation (Chapter 4), to obtain a prediction of such vertical velocities solely as a function of the wind stress field!

Let's begin with a reminder of the continuity equation, rearranged with a head start toward solving for vertical velocities:

$$-\frac{\partial w}{\partial z} = \frac{\partial u}{\partial x} + \frac{\partial v}{\partial y} \tag{6.4}$$

If we integrate both sides of Equation 6.4 from the surface to z_{Ek}, then we again eliminate the vertical partial derivative (from the left-hand side) and the horizontal velocities u and v on the right-hand side are, by definition, the Ekman transports. The resulting expression,

$$w_{Ek} = \frac{\partial}{\partial x}(U_{Ek}) + \frac{\partial}{\partial y}(V_{Ek}) \tag{6.5}$$

says quite simply that one will find positive vertical (upward) velocity where the Ekman transports are diverging, and downward velocity where the Ekman transports are converging. It only makes sense; these are natural consequences of the incompressibility of water and the fact that there cannot be a vacuum without being filled by adjacent water. We call the resulting vertical velocity **Ekman pumping velocity** (w_{Ek}) and, by convention, refer to upward velocity ($w_{Ek} > 0$) as Ekman suction and downward velocity ($w_{Ek} < 0$) as Ekman pumping.

Since our Ekman pumping velocity is, thus far, expressed as a function of Ekman transports, and we have already derived an expression for Ekman transport as a function of wind stress (Equation 6.3), let us now close the loop and substitute those into Equation 6.5:

$$w_{Ek} = \frac{\partial}{\partial x}\left(\frac{\tau_y}{\rho f}\right) + \frac{\partial}{\partial y}\left(\frac{-\tau_x}{\rho f}\right) \tag{6.6}$$

Within a reasonable radius, we can consider ρ and f constant, and pull them out of the horizontal derivatives:

$$w_{Ek} = \frac{1}{\rho f}\left(\frac{\partial \tau_y}{\partial x} - \frac{\partial \tau_x}{\partial y}\right) \tag{6.7}$$

which might be recognizable from the previous chapter as nothing but the *wind stress curl* divided by ρf, or

$$w_{Ek} = \frac{\nabla \times \tau}{\rho f} \tag{6.8}$$

How fortunate! Obtaining actual measurements of Ekman transports is very challenging, whereas estimates of the global surface wind field are now routinely made by satellites and other methods, including atmospheric models that ingest global (albeit sparse over the ocean) weather observations – a process called data assimilation. Thanks to **V.** Walfrid Ekman and the continuity equation, we can diagnose or predict vertical motions in the ocean – something that is virtually impossible to measure yet has crucial implications ranging from local biology to global climatology – simply by calculating the curl of the wind stress field!

Positive wind stress curl equates to upwelling in the Northern Hemisphere – so says the mathematics, but let's check that it squares with our physical intuition. A nonzero wind stress curl can arise from zonal shear of meridional wind stress ($\partial\tau_y/\partial x$) or from meridional shear of zonal wind stress ($\partial\tau_x/\partial y$). Three different, simplified scenarios are depicted in Figure 6.2, each yielding positive vertical velocity by way of Ekman dynamics. In Figure 6.2a there is no zonal wind stress, but the meridional wind stress is arranged such that the Ekman transports (directed 90° to the right of each wind stress vector) are diverging away from the central region, which would induce, by continuity, rising water to take its place. In this case, $\partial\tau_y/\partial x > 0$ so $w_{Ek} > 0$. In Figure 6.2b there is no meridional wind stress, but the zonal wind stress reverses direction over some latitude and, applying the same general rule, the Ekman transports are again directed away from one another. In this case, $\partial\tau_x/\partial y < 0$ and, since the expression for Ekman pumping velocity includes a negative sign before that term (Equation 6.7), the result is once again $w_{Ek} > 0$. Finally, the wind stress pattern portrayed in Figure 6.2c, which might resemble that of a tropical cyclone in the Northern Hemisphere (counterclockwise or cyclonic rotation), has both terms contributing to a positive wind stress curl. At any point along the wind circulation, the Ekman transport will be directed outward from the center, so there must be upwelling in the center. Confirming this intuition with Equation 6.7, $\partial\tau_y/\partial x > 0$ and $\partial\tau_x/\partial y < 0$ in the center of the circulation, so $w_{Ek} > 0$ – and indeed upwelling can be found beneath such broad cyclonic circulations. Keep this in mind. It is not the wind stress curl *per se* that *drives* Ekman pumping, but the curl provides a convenient *shortcut* to it by determining whether the wind stress field is arranged such that Ekman transports are *diverging* or *converging*, since one is always perpendicular to the other.

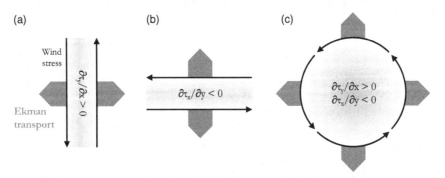

Figure 6.2 Schematic illustration of wind stress and resulting Ekman transports leading to positive vertical velocity (Ekman suction) for three cases in the Northern Hemisphere.

6.2.2   Wind-Driven Upwelling Zones

Some of the most consequential examples of diverging Ekman transports, or Ekman suction, in the world ocean are arguably along *boundaries*. One such type of boundary is that between land and sea, where not only are there often strong horizontal gradients in wind speed, but any Ekman transport not directed exactly parallel to the coastline will, by definition, result in a horizontal convergence or divergence thereof. Take the west coast of the USA in the northeastern Pacific Ocean, for example (Figure 6.3). Recall from Chapter 5 that this is where we should find the eastern half of the subtropical high over the North Pacific, which of course comes with northerly winds running along the coastlines of the US states of Washington, Oregon, and California. Equipped with the dynamics outlined in the previous section, we already have enough information to say quite a bit about the three-dimensional ocean circulation we would expect to find in this region.

The simplest way to proceed is to infer from the definition of Ekman transports that the upper layer of water (down to z_{Ek}) must be moving 90° to the right of the wind stress, which in this case means flowing straight away from the coast. This is clearly a horizontal *divergence*, because the water is flowing away from a boundary across which there is no more water (the beach). Continuity requires that this horizontal transport divergence be balanced by a vertical convergence. Upwelling thus fills the hypothetical void left behind by the offshore-directed Ekman transport – that is, $w_{Ek} > 0$ along the coast. This wind-driven upwelling can also be understood in terms of Ekman pumping and its constituent wind stress curl. While

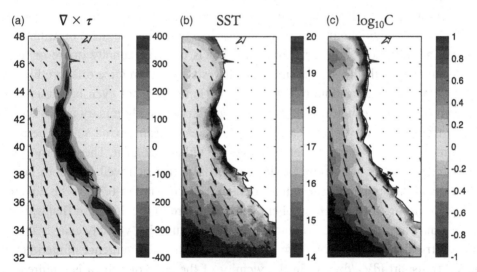

Figure 6.3 Wind stress curl (a), sea surface temperature (b), and the logarithm (base 10) of chlorophyll concentration (c) in the northeastern Pacific Ocean (along the west coast of the USA). In each panel, the wind stress is also shown. Data are from the boreal summer climatology (June–August) from the following satellite products: CCMP v.2 winds and MODIS Aqua. (A black and white version of this figure will appear in some formats. For the color version, please refer to the plate section.)

the reach of the aforementioned northerly winds associated with the atmospheric general circulation certainly extends some distance onto land, they are dramatically weakened there due to the presence of significant friction and orographic features. Along the coast, then, there is a very large, positive $\partial \tau_y / \partial x$ (shown in Figure 6.3a), which equates to Ekman suction ($w_{Ek} > 0$) by Equation 6.7.

The manifestations of Ekman suction associated with this and other wind-driven, coastal upwelling zones can be readily observed from space using modern satellite measurements. Almost perfectly aligned with the thin, coastal strip of positive wind stress curl is a similar strip of cold SST. This is because Ekman pumping velocity can connect directly to the ocean mixed layer heat budget by way of the vertical advection term ($-w\, \partial T / \partial z$). In this case, the upwelling ($w_{Ek} > 0$) combined with typical stratification ($\partial T / \partial z > 0$) results in cold temperature advection ($-w_{Ek}\, \partial T / \partial z < 0$). Surfing apparel is not the only thing influenced by the ~14 °C summertime SST. Chlorophyll is a pigment produced by photosynthesizing organisms – the water's greenness, if you will – that certain satellite instruments are designed to detect and provide a proxy of biological productivity. Along the same coastlines, such "ocean color" sensors detect extremely high concentrations of chlorophyll (Figure 6.3c). While there are a great many factors (physical, biological, and chemical) that determine photosynthesis rates and overall productivity, the general pattern can be understood similarly just by swapping nutrients for temperature in the vertical advection term. Upwelling brings water with higher concentrations of nutrients from depth into the sunlit surface zone, where they can be utilized by phytoplankton, generating extremely high surface chlorophyll concentrations.

Upwelling zones with similar intensity and consequence are found in both hemispheres, and along boundaries oriented in any direction. The continent of Antarctica provides a continuous land–sea boundary that is oriented zonally (more or less), which interacts with the strong, unimpeded westerlies to produce offshore Ekman transport and upwelling (Figure 6.4). However, land need not even be present! In the open ocean, the equator itself can effectively serve as a boundary across which Ekman transports may oppose each another under the same wind simply because the sign of the Coriolis parameter f changes on either side of it. The easterly trade winds in the tropics, which are "wide" enough (Δy) to reach well into both hemispheres, will cause meridional Ekman transport V_{Ek} (Equation 6.3b) to be directed northward in the Northern Hemisphere and southward in the Southern Hemisphere. While it may be problematic to compute the Ekman pumping velocity exactly *on* the equator (try dividing by zero), there is clearly a meridional divergence of meridional Ekman transport ($\partial V_{Ek} / \partial y > 0$) in the vicinity of the equator, which is required by continuity to be balanced by upwelling. In all cases, there are consequences for SST (carrying the possibility of an atmospheric response) and biogeochemical processes.

The equatorial Pacific Ocean is the largest natural source of carbon dioxide (CO_2) to the atmosphere. This fact is a real testament to the significance of

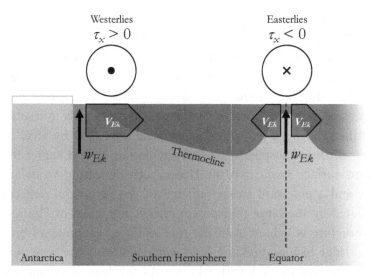

Figure 6.4 Schematic illustration of wind-driven upwelling along the coast of Antarctica in the Southern Hemisphere, and along the equator. In both cases, there is a meridional divergence of meridional Ekman transport (V_{Ek}), resulting in Ekman suction ($w_{Ek} > 0$). Circles with a dot (cross) represent wind directed out of (into) the page.

the wind-driven equatorial upwelling, because the vertical transport of carbon to the surface actually overwhelms two other influences that would otherwise tilt the equatorial Pacific toward being a sink. One is that there is a great deal of photosynthesis (due to upwelling of nutrients), which consumes carbon, lowering the surface concentration of CO_2, and would thus tend to draw CO_2 down from the atmosphere. Additionally, the coolness of the surface directly on the equator (again, due to upwelling) adds a tendency for CO_2 to be drawn down since its solubility is inversely proportional to the solvent's temperature. The equatorial Pacific Ocean and its crucial role in climate variability, both natural and anthropogenic, is a centerpiece of the next chapter.

6.2.3   The Subtropical Gyres

We now aim our growing set of tools and rules at understanding a general feature of the wind-driven circulation that dominates fully one-quarter of the surface area of Earth: the great subtropical ocean gyres. Not only are the "gyres" of great importance in terms of the global ocean circulation and climate, but they serve as a rigorous test of our understanding of Ekman dynamics, geostrophy, and the heat budget. Imagine a simple wind circulation over one of the subtropical ocean basins – the North Atlantic, for instance – that is a perfectly symmetric circle flowing clockwise or anticyclonic. (Of course, we know from Chapter 5 that the subtropical highs are not perfectly symmetric, but let's assume they are for this hypothetical question.) Beneath it, *the ocean circulation also flows clockwise. Why is this?* To the uninitiated, it may seem as simple as the atmosphere is dragging the ocean in the same direction. But the connection is so much richer than that, and you now have everything you need to explain it. Further details, such as the fact that the ocean gyres are *not* symmetrical (and they would remain asymmetrical even if the subtropical highs *were* symmetrical), were worked out by Ekman's successors like Sverdrup, Stommel, and Munk. From our vantage point – fortunately one of hindsight – even those details don't require mathematical cartwheels to resolve.

Without further ado, let's build that subtropical gyre step by step. The oversimplified view of the atmospheric general circulation (from Chapter 5) includes an equatorward surface branch of the Hadley cell deflecting westward by the Coriolis force, and a poleward surface branch of the Ferrel cell deflecting eastward, as indicated by the large transparent arrows in step A of Figure 6.5. This anticyclonic structure is enhanced in the real world by landmasses breaking up the subtropical ridge of high pressure (that would otherwise be wrapping around the planet) with quasi-stationary thermal lows. All this leads to an impressive and resilient anticyclonic circulation over the major subtropical ocean basins with a signature negative wind stress curl (step A). The inferred Ekman transports based on the given wind field are everywhere inward (step B). (Such convergence of Ekman transports occurring in the Southern Hemisphere can also be visualized in Figure 6.4.) The converging Ekman transports result in Ekman pumping ($w_{Ek} < 0$) by continuity (step C), which could have been arrived at without step B simply by applying Equation 6.8. This Ekman pumping velocity operates through the vertical advection term in the heat budget to warm the original location of the thermocline (i.e., where there is a strong $\partial T/\partial z$), or deepen the thermocline, effectively

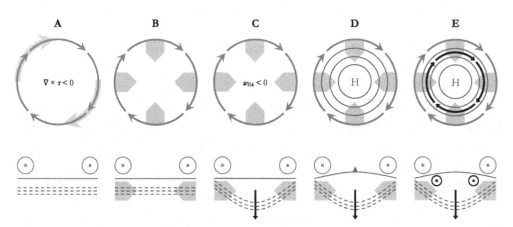

Figure 6.5 Schematic illustration of the mechanisms involved in the wind forcing of the subtropical ocean gyres, valid for the Northern Hemisphere. Wind in blue, sea surface height in solid red, subsurface isotherms (thermocline) in dashed red, Ekman transports in gray, Ekman pumping velocity in black, and geostrophic current in black. The thick transparent arrows (shown only in step A) represent the surface branches of the Hadley and Ferrel cells. (A black and white version of this figure will appear in some formats. For the color version, please refer to the plate section.)

increasing the average temperature of the water column. From Charles' Law (thermal expansion), the warmer layer of water consumes more volume, which raises the sea level like a mirror image of the depressed thermocline (step D). The final step in building our model wind-driven gyre is simply to infer the geostrophic flow from the sea surface height field, just as we did in Chapter 4. Since the pressure gradient force is directed outward from the center of the sea level hill, the geostrophic balance will be achieved when the velocity is flowing clockwise, parallel to the sea surface topography lines (step E). Comparing our final product with our starting point, we indeed have a clockwise wind pattern driving a clockwise gyre, and we needed almost our entire dynamical toolbox to say why!

Before making some real-world refinements to our idealized wind-driven gyre circulation, it is worth acknowledging some of the scales involved in setting this enormous, globally important piece of the ocean circulation that virtually dominates each ocean basin. If your boat was sitting at the very summit of the sea level hill somewhere in the middle of one of the subtropical ocean basins, you would be about 1 m higher than the base of the hill *thousands of kilometers away*. Such slopes obviously cannot be perceived by the naked eye, and frankly it is amazing that we have engineered the technology to detect them using altimeters in Earth orbit. One meter – that's what it takes to set the whole thing in motion.

Slightly more mysterious is that these great subtropical ocean gyres are anything but zonally symmetric. Their anticyclonic circulation (clockwise in the Northern Hemisphere, counterclockwise in the Southern Hemisphere) is heavily skewed toward the west – a behavior known as **western intensification**. The resulting intense, poleward-flowing currents along the western boundaries of the subtropical oceans are known as **western boundary currents**, with the Gulf Stream and Kuroshio Currents being examples. To understand them ourselves, let's return to the twentieth century to follow the logic of another Scandinavian oceanographer, Harald Sverdrup (1888–1957). Sverdrup endeavored to expand upon Ekman's linear, steady-state framework, but without neglecting the pressure gradient force in the horizontal momentum budget (Sverdrup, 1947). Sverdrup was also not interested in diagnosing vertical motions in his landmark 1947 paper, so he made an assumption of horizontal nondivergence, yielding the following equations of motion as a starting point:

$$\frac{1}{\rho}\frac{\partial p}{\partial x} = fv + \frac{1}{\rho}\frac{\partial \tau_x}{\partial z} \tag{6.9a}$$

$$\frac{1}{\rho}\frac{\partial p}{\partial y} = -fu + \frac{1}{\rho}\frac{\partial \tau_y}{\partial z} \tag{6.9b}$$

$$\frac{\partial u}{\partial x} + \frac{\partial v}{\partial y} = 0 \tag{6.9c}$$

Sverdrup followed in Ekman's footsteps by taking the vertical integral of the remaining terms in the horizontal momentum budget, but first multiplying through by density ρ so the resulting transports U and V are formally **mass transport** (with units kg/s), yielding

$$\frac{\partial P}{\partial x} = fV + \tau_x \tag{6.10a}$$

$$\frac{\partial P}{\partial y} = -fU + \tau_y \tag{6.10b}$$

where pressure P is capitalized to denote the vertically integrated pressure. Then, in a stroke of mathematical poetry, Harald cross-differentiated Equation 6.10 and invoked Equation 6.9c, yielding one of the most famous (if not memorable) expressions in physical oceanography:

$$\frac{\partial f}{\partial y} V = \frac{\partial \tau_y}{\partial x} - \frac{\partial \tau_x}{\partial y} \tag{6.11a}$$

$$\beta V = \nabla \times \tau \tag{6.11b}$$

where β represents the dependence of the Coriolis frequency on latitude $\partial f / \partial y$. Recognizing that the $\partial \tau_y / \partial x$ component of the wind stress curl is very small compared to $\partial \tau_x / \partial y$ in the real world (which you can rationalize by examining the figures in Section 5.4.3), Sverdrup further simplified Equation 6.11 to

$$\beta V = -\frac{\partial \tau_x}{\partial y} \tag{6.12}$$

The resulting expression is known as the **Sverdrup balance**, and states that the meridional mass transport V in the open ocean will be governed, to first order, by the meridional shear of the zonal wind stress $\partial \tau_x / \partial y$. The Sverdrup balance has a remarkable applicability to the real world, since between about 20–50° latitude in either hemisphere, the zonal wind stress transitions from the tropical easterlies to the midlatitude westerlies (see again the figures in Section 5.4.3). Therefore, the Sverdrup balance predicts that there ought to be equatorward mass transport everywhere in the subtropics. It is true that there is a relatively weak equatorward flow extending across *most* of the subtropical ocean basins, but continuity requires a return flow, and since we've already established the direction of the gyre circulation (anticyclonic, like the overlying wind field), the only place for that to occur is along the western boundary. It is interesting that Sverdrup balance does not quite predict western intensification; it predicts that most of the basin should exhibit equatorward flow (generally referred to as *eastern* boundary currents) and only by corollary (or mass continuity) must there be narrow yet intense poleward-flowing *western* boundary currents.

The one to whom credit for explaining western intensification really goes is Henry Stommel (1920–1992). In his landmark paper, published just a year after

Sverdrup's, Stommel showed us that the *key* is that the magnitude of the Coriolis frequency *f* varies with latitude (Stommel, 1948). Coincidentally, Sverdrup's namesake balance even has an explicit recognition of this fact ($\beta = \partial f / \partial y$), but he had applications to other currents in mind at the time. Stommel's discovery that the latitudinal variation of the Coriolis frequency leads to western intensification of the gyres is best understood through the conservation of **potential vorticity**. Around 1940, Swedish-born meteorologist Carl-Gustaf Rossby (1898–1957) coined the term potential vorticity (or PV for short) and argued that it is conserved in a Lagrangian sense – that is, the *total* PV of a fluid parcel won't change as it moves around in the ocean or atmosphere (Rossby, 1940):

$$\frac{\mathrm{D}}{\mathrm{D}t}\left(\frac{\zeta + f}{H}\right) = 0 \tag{6.13}$$

where ζ is **relative vorticity** (but that's just *curl* of the ocean velocity field!), *f* is **planetary vorticity** (and of course, that's just the Coriolis frequency), *H* is the depth or thickness of the fluid layer, and recall the capitalized D indicates the total (or Lagrangian) derivative, in contrast to ∂ for partial (or Eulerian) derivatives. All these vorticities and how they are going to balance each other can seem overwhelming, so let's summarize before proceeding further (Table 6.1).

There are actually only two *unique* vorticity constructs – ζ and *f* – while the others – η and PV – are just combinations thereof. For our purposes, we will consider *H* constant, so our application of PV to the task of understanding western intensification will only involve two moving parts – the tradeoff between relative and planetary vorticity needed to prevent PV from changing. Let's also keep in mind that relative vorticity (aka curl) is just the tendency of the flow field to cause a fluid parcel contained therein to spin. Imagine (or sketch) a flow field that is rotating in a counterclockwise direction (which is cyclonic in the Northern Hemisphere). Such a flow field will have meridional velocity heading in opposite directions on the west and east ends (such that $\partial v / \partial x$ is positive) and zonal velocity heading in opposite directions on the north and south ends (such that $\partial u / \partial y$ is negative), rendering the curl of the velocity field, or the relative vorticity ζ of a parcel therein, positive.

Table 6.1 A summary of the vorticities discussed in the text

Symbol	Name	Other name	Equation
ζ	Relative vorticity	Curl	$\partial v / \partial x - \partial u / \partial y$
f	Planetary vorticity	Coriolis frequency	$2\Omega \sin \varphi$
η	Absolute vorticity		$\zeta + f$
PV	Potential vorticity		$(\zeta + f)/H$

Let us now take a ride on a poleward-flowing parcel on the western side of one of the great subtropical ocean gyres (Figure 6.6). For this thought experiment, we can simply resume where we left off after building the symmetrical gyres (step E of Figure 6.5). As our parcel is moving northward, its planetary vorticity f is increasing, since the Coriolis frequency is a function of latitude. Given the form of potential vorticity (Equation 6.13), and with H constant, the only way to conserve PV is to balance the growing f by decreasing the relative vorticity ζ by the same amount. What is it to decrease ζ but to impart a clockwise spin (left-hand inset in Figure 6.6)? The sense of the rotation induced through the conservation of PV is such that some northward flow is added along its western side and subtracted from its eastern side. Now, we can repeat this process for the newly spawned northward flow on the western side of the parcel; it sees increasing f as it flows northward, which is balanced by further decreasing ζ. Consider this process repeating itself with each successive poleward flow induced to the immediate west of each previous parcel, ultimately intensifying the poleward flow to the west and weakening it to the east until reaching the continental coastlines defining the western boundaries of the subtropical oceans.

The result of an otherwise symmetric gyre under the constraint of PV conservation is therefore a very asymmetric gyre such that the poleward flow along the western boundary is fast and narrow while the equatorward flow can be weak yet broad (achieving the same total transport). This is indeed the picture painted by the observed, time-averaged near-surface velocity and sea surface height fields (see Section 4.3 on geostrophic balance). It is quite clear why Stommel emphasized that the latitudinal variation of f is critical for explaining the western boundary currents; if f was a constant, there would be no need for ζ to adjust and intensify

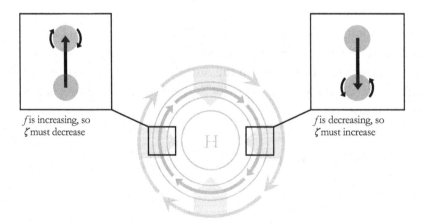

f is increasing, so
ζ must decrease

f is decreasing, so
ζ must increase

Figure 6.6 Schematic illustration of the conservation of potential vorticity amid meridional flows on either side of a subtropical ocean gyre, valid for the Northern Hemisphere. The faded element in the center of the illustration is step E from Figure 6.5. (A black and white version of this figure will appear in some formats. For the color version, please refer to the plate section.)

(weaken) the poleward flow on the western (eastern) sides of the gyres. Stommel and his contemporaries, particularly Walter Munk (1917–2019), refined this view of western intensification even further by accounting for friction along the sea-floor and along the continental boundaries (via horizontal eddy diffusion), and even the geometry of the subtropical ocean basins being more like triangles than rectangles (Munk and Carrier, 1950), but we can leave those details to a dedicated

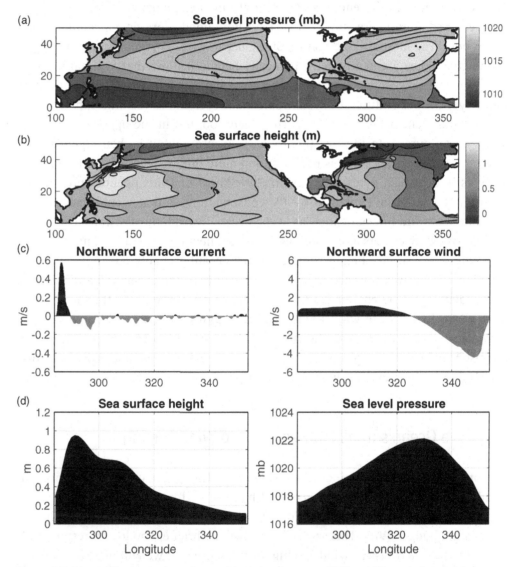

Figure 6.7 Maps of long-term annual mean (a) sea level pressure (SLP) and (b) sea surface height (SSH) in the North Pacific and Atlantic. (c) Profiles of meridional surface velocity in the ocean (left) and atmosphere (right) along ~35°N in the North Atlantic. (d) As in (c), but profiles of SSH (left) and SLP (right). SLP data from the NOAA/NCEP-DOE Reanalysis 2 data set (2.5°, 1988–2017), SSH data from the AVISO product (1°, 1993–2010), surface currents from the OSCAR product (1/3°, 1993–2010) and surface winds from CCMP v.2 product (1/4°, 1988–2017). (A black and white version of this figure will appear in some formats. For the color version, please refer to the plate section.)

course in geophysical fluid dynamics. In the following section, though, we will discuss a couple of types of waves that propagate within the ocean and are vital for climate variability. Just to foreshadow, the westward drift of momentum as just described, which is due to the *variation of Coriolis frequency with latitude*, is equivalent to what we appropriately call a Rossby wave.

Before leaving the gyres behind for a while, it is worth taking a step back to ponder the similarities, differences, and potential *interactions* between the "gyres" of the ocean and atmosphere, now that we have built good, working conceptual models for both. The most striking feature of the ocean gyres is arguably their extreme asymmetry – they are nearly geostrophically balanced flows that *lean* hard to the west, and we now understand why. The observant student may have noticed a similar yet curious feature of the subtropical anticyclones in the atmosphere, from the previous chapter (Section 5.4.2). They lean, too, but in the opposite direction to the ocean gyres – to the east (Figure 6.7). Accordingly, it is their equatorward flow on the eastern sides that is narrower and more intense (even if not as dramatically as in the ocean gyres). *Why is this?* We know that this phenomenon in the atmosphere cannot be explained by PV conservation, as that would have led to the same westward intensification as in the ocean. Instead, the eastward leaning of the high-pressure centers and bunching of isobars near the coastlines (analogous to bunching of sea surface height contours in the ocean) probably has to do with the coupling, or two-way interaction, between the gyres of the ocean and atmosphere. The student is hereby challenged to synthesize material presented thus far in this textbook to hypothesize why the atmospheric "gyres" (i.e., the subtropical highs shown in Figure 6.7a) *lean* to the right, in clear opposition to the leaning of the great subtropical ocean gyres. Don't forget about coastal upwelling, nor about the last section in the previous chapter on atmospheric responses to SST variability!

6.3 The Ocean's Transient Response to Wind Forcing

The previous section examined the response of the upper ocean circulation to steady winds. That is, the time-*averaged* balance of forces in which we always made the steady state approximation by neglecting the time derivative in the momentum budget $\partial \mathbf{V}/\partial t$. This section will give one brief example of a response of the ocean to *true* transient wind forcing, in the sense that we will need to account for the time derivatives to understand the flow, but then we will move toward a transient response that might feel a little bit steady. This is because we will begin with some dynamics familiar from earlier in this chapter, while the winds are "switched" on just long enough for, say, Ekman transports to emerge, but we can still consider it a transient response because it is only after those winds are "switched" off that the ocean's response shines.

6.3.1 Inertial Oscillations

If you've become convinced by the steady-state response of the ocean to wind forcing, like the beautiful Ekman spiral, you will quickly be puzzled as an observational oceanographer in the field. Superimposed upon the steady-state wind field, on any given day, is a mess of effectively random (and transient) weather. The atmospheric *general* circulation is, in effect, buried amid all that noise. What you are more likely to measure, if you find yourself in the open ocean searching for the wind-driven ocean circulation, is a cacophony of small-scale ocean currents driven by wind fluctuations that last minutes to hours. Such wind forcing is sufficiently brief (even if intense) that the resulting currents, for most of their life cycle, are powered simply by their remaining *inertia*, so we call them **inertial oscillations**.

Let's conduct another simple thought experiment, with a careful accounting of the horizontal momentum budget. Let's start with a very boring oceanographic situation: the ocean is flat ($\nabla \eta = 0$) and motionless ($u = v = 0$), so our horizontal momentum budget reduces to nothing. Now let's switch on a brisk southerly wind (positive τ_y), but just for about one hour. Now we have the following momentum budget:

$$\frac{\partial u}{\partial t} = 0 \tag{6.14a}$$

$$\frac{\partial v}{\partial t} = \frac{1}{\rho}\frac{\partial \tau_y}{\partial z} \tag{6.14b}$$

Based on Equation 6.14b, we can see that the moment the southerly winds are switched on, the water beneath them will accelerate northward, and there is not yet any reason for there to be a zonal component of the ocean velocity. There is also not enough time for the horizontal momentum to be transferred very far downward into the water column as stress, so the Ekman spiral and transports are not going to form in time. By the time the winds are switched off (i.e., eliminate the right-hand side of Equation 6.14b), there has only been some northward velocity v generated very near the surface, so we have

$$\frac{\partial u}{\partial t} = fv \tag{6.15a}$$

$$\frac{\partial v}{\partial t} = 0 \tag{6.15b}$$

where the Coriolis term fv now appears in the zonal momentum budget given the nonzero v. From the zonal momentum budget (Equation 6.15a), we can see that there must now be a positive zonal (eastward) acceleration of the water. Once that has had a little time to come to fruition, both the meridional *and* zonal components of the velocity (or momentum) are nonzero, so, finally, we have

$$\frac{\partial u}{\partial t} = fv \tag{6.16a}$$

$$\frac{\partial v}{\partial t} = -fu \tag{6.16b}$$

It might not take long to realize that these two equations are linked to one another and constitute the perfect recipe for a circle (hence the term "oscillation" in their name). In other words, the more northward flow, the more it turns eastward, and the more eastward flow, the more it turns southward. Around and around we could go.

The above thought experiment seems to imply that every time a gust of wind blows over the ocean for a short period of time (say, less than a day), perpetual and perfectly circular oscillations will be spawned. Of course, they are not perpetual because they have no source of momentum but their inertia, which will be eroded over time by frictional dissipation – horizontal and vertical eddy diffusivity, for example. Neither will they be perfectly circular. Recall from the previous section that in discussing the PV balance leading to western intensification of the gyres, our inertial oscillations will be subject to the same laws and also tend to drift westward – *and* wherever the larger-scale eddies and perhaps even steady-state currents in the background sweep them.

An exquisite example of inertial oscillations was observed in 2018 as physical oceanographers Elizabeth Sanabia and Steven Jayne dropped a series of profiling floats[1] in a straight line out of an airplane, just ahead of Hurricane Florence in the tropical North Atlantic Ocean. Floats deployed where the southern half of Florence passed over traveled eastward in the long run, and floats deployed beneath the northern half of Florence traveled overall westward – just like the counterclockwise winds of the hurricane itself (Figure 6.8). However, embedded within those

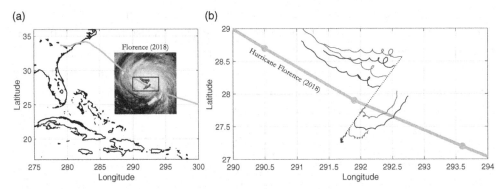

Figure 6.8 Inertial oscillations generated by Hurricane Florence in 2018. The overview map (a) shows the path and scale of Florence relative to the deployed floats. The inset map (b) shows the detailed trajectories of the floats during and after the passage of Florence. Image of Florence taken by NOAA/GOES-16 satellite and processed by the Cooperative Institute for Research in the Atmosphere (CIRA). Data from the Air Launched Autonomous Micro Observer (ALAMO) float program courtesy of Elizabeth Sanabia, US Naval Academy and Steven Jayne, Woods Hole Oceanographic Institution. (A black and white version of this figure will appear in some formats. For the color version, please refer to the plate section.)

[1] The floats deployed just ahead of the path of Hurricane Florence (2018) were Air Launched Autonomous Micro Observer (ALAMO) floats. ALAMO floats are similar to, but smaller than, the Argo floats introduced in Chapter 3 (Jayne and Bogue, 2017). For more information including scientific publications, visit the ALAMO program home page hosted by the Woods Hole Oceanographic Institution (WHOI): https://alamo.whoi.edu/

general paths are clockwise meanders that decay over time – classic inertial oscillations thrust into motion by the brief but very intense winds of a major-category hurricane. The broader climatic importance of inertial oscillations likely involves a role in horizontal mixing. They are also an excellent starting point for understanding two types of waves that are unquestionably important in climate dynamics.

6.3.2 Kelvin and Rossby Waves

Let's repeat the above thought experiment, but this time allow our modestly sized patch of wind stress to be switched on just long enough for some of the key elements of the steady-state response to form, particularly Ekman transport. Let's call it an even 10 days. Let's also make it interesting, and set our wind patch directly over the equator, and this time have it blow from west to east. Such a case is illustrated in step A of Figure 6.9.

In the Northern Hemisphere, the Ekman transport due to such a westerly wind anomaly will directed southward, and in the Southern Hemisphere it will be directed northward – in both cases, toward the equator and thus ensues a convergence of Ekman transports (step B). By continuity, we know that a convergence of Ekman transports induces Ekman pumping ($w_{Ek} < 0$) and, as in the case of the subtropical gyres, will result in a depressed thermocline ($-w_{Ek} \, \partial T/\partial z > 0$) to be mirrored on the sea surface. Not only will there be a high sea surface height anomaly on the equator, but the two locations *from* which water was transported will have *lower* sea level resulting from the Ekman *divergence* there (step C).

Now we can switch off our imposed patch of westerly wind, but we are left with one hill and a pair of valleys on the sea surface, to which the ocean must now adjust. One way that the ocean (or any fluid medium) adjusts to such unsustained mass imbalances (or potential energy) is to dissipate them by *waves*. Let's try to rationalize, using intuitive dynamics rather than just believing so-called wavelike solutions to the equations, why we observe the resulting sea level anomaly on the equator (which we'll call an equatorial **Kelvin wave**) to always propagate eastward and the resulting pair of sea level anomalies *off* the equator (which we'll call **Rossby waves**) to always propagate westward. The geostrophic velocity field about the high sea level anomaly on the equator is such that it will diverge from the west

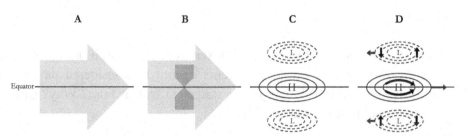

Figure 6.9 Schematic illustration of the production of Kelvin and Rossby waves by a temporary westerly wind stress centered on the equator. (A black and white version of this figure will appear in some formats. For the color version, please refer to the plate section.)

side of the hill (on the "back end") and converge upon the east side of the hill (the "front end"). This has the effect of progressively lowering sea level on the west and raising sea level on the east, which is to say, propagating the essence of the feature eastward. Traveling eastward along with it will be the full complement of high sea level anomaly, downwelling beneath and generally eastward geostrophic currents – all symmetric about the equator. One could also note that eastward is the *only* option for a Kelvin wave, as defined by a wave where the Coriolis force is balancing a boundary, since westward propagation would make such a balance impossible (conjuring a vision of the excess mass splitting away from the equator into both hemispheres, consistent with the 90° right/90° left mantra).

Why do the sea level depressions found off the equator propagate westward as Rossby waves? They do so for the same reason that the subtropical gyres lean westward and inertial oscillations drift westward: potential vorticity is conserved. For example, a parcel of seawater following the equatorward (geostrophic) flow on the west side of the low sea level anomaly in the Northern Hemisphere is experiencing a reduction of planetary vorticity f, so it must undergo an increase in relative vorticity ζ in order to conserve PV. An increase in relative vorticity means acquiring a spin such that there is equatorward flow added to the west side of the parcel and subtracted from the east side, so the signal propagates westward. On either side of the low sea level anomalies, in either hemisphere, one will find the balancing between f and ζ leading to a westward propagation of the anomaly.

While the above thought experiment began with a zonal wind stress anomaly on the equator, Kelvin waves also use coastlines as the boundary against which the Coriolis force is balanced (known as coastally trapped Kelvin waves). Similar to the equatorial variety, coastal Kelvin waves also only have one possible direction of propagation. Rossby waves are even more universal. While they always propagate westward in the ocean, they can emerge at any latitude, near or away from the coast, in response to wind fluctuations and other phenomena capable of setting up (and then leaving behind) a mass imbalance. In the next chapter, Kelvin and Rossby waves will reappear to play a central role in the most dominant form of year-to-year climate viability on Earth.

Questions

1. The gradient of sea surface height $\nabla\eta$ serves as the engine for geostrophic currents. Therefore, geostrophic currents cannot be considered part of the wind-driven ocean circulation. Is this statement true or false? Why? Give an example in support of your answer.

2. How is the Ekman layer depth (z_{Ek}) defined?
3. If a steady wind in the Northern Hemisphere blows from north to south (i.e., northerly), toward which direction will the upper ocean (say, averaged over the top 100 m) flow due to Ekman transport?
4. Marked on the map is an imaginary region of cold SST along the southeastern coast of South America. Based on your understanding of Ekman dynamics, which direction should the steady wind be blowing, and why?

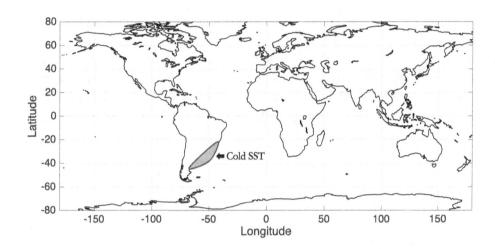

5. Negative SST anomalies are often observed immediately after the passage of a strong hurricane (a so-called cold wake). Using any concept(s) from Chapters 2, 4, or 6, develop a plausible mechanism by which a hurricane could cool the ocean surface.
6. Sketch two different surface wind circulation patterns that would result in Ekman suction ($w_{Ek} > 0$), one for the Northern Hemisphere and one for the Southern Hemisphere. You may invoke a lateral boundary (i.e., a coastline) in one case. Next to both cases, write the Ekman pumping velocity equation ($w_{Ek} = ...$) with the wind stress curl term fully expanded into its component parts and identify exactly why $w_{Ek} > 0$ for both cases.
7. Consider the latitudinal profile of zonal wind stress τ_x shown in the following figure. Reproduce the figure including the two blank graphs for $\partial \tau_x / \partial y$ and meridional mass transport V, and complete them based on Sverdrup balance. Do not be concerned with units in these sketches. Identify the latitude band(s) in which Sverdrup balance predicts northward mass transport.

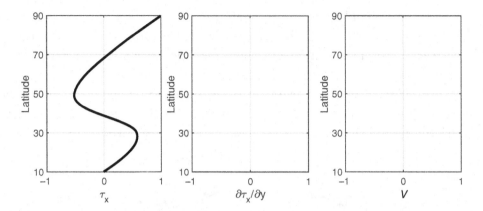

8. Consider the latitudinal profile of meridional mass transport V given in the following figure; note that the domain is in the Southern Hemisphere. Reproduce the figure, including the blank graph for τ_x, and complete it by sketching a latitudinal profile of τ_x consistent with the V profile based on Sverdrup balance. Do not be concerned with units.

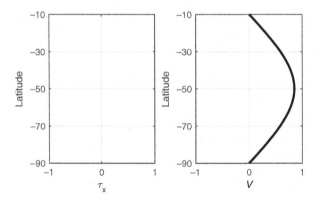

9. Describe, step-by-step and with a potential vorticity perspective, the process by which a *negative* sea surface height anomaly in the Southern Hemisphere (at least a few degrees away from the equator) propagates westward. A sketch may be given to supplement your explanation.

10. Briefly explain why the great subtropical ocean gyres flow in the same direction (e.g., clockwise in the Northern Hemisphere) as the general large-scale wind pattern above.

11. Exactly what is it about the Coriolis force that enables the gyres to be shaped so asymmetrically relative to the overlying general large-scale wind pattern?

12. Compare and contrast Ekman transport and inertial oscillations in terms of the assumptions and simplifications made to the momentum budget to describe them mathematically.

13. Why is the diameter of an inertial oscillation smaller at high latitudes than one with the same initial velocity perturbation at low latitudes?

14. Consider an easterly (from east to west) wind stress anomaly, centered on the equator in the western equatorial Atlantic Ocean. (a) What is the name of the oceanic wave that will likely propagate eastward from that location? (b) Will it cause a warming or cooling of SST in the eastern equatorial Atlantic after some time, and why?

15. Select one of the *Dive into the Data* boxes featured in this chapter. Read the sample journal article that uses the associated data set, and describe the general role of that data set in the study. Provide a little context, both technical and scientific. What did the authors actually "do" with the data set? What specific scientific insight was enabled by incorporating this data set into the study?

16. Utilizing *Dive into the Data* Box 4.3, produce a Hovmöller diagram (longitude by time) of monthly sea surface height (η) anomalies along the equator and examine the equatorial Kelvin waves in the Pacific basin. (a) Identify one Kelvin wave; note the approximate date and longitude at which it first appears in the western equatorial Pacific, and estimate its phase speed in m/s based on the time it took to propagate from the western to eastern equatorial Pacific Ocean. (b) Do you think this data set, as provided, is ideal for examining equatorial Kelvin waves? If not, specify any limitations you think prevent it from fully resolving Kelvin waves and their propagation characteristics.

17. Utilizing *Dive into the Data* Boxes 6.1, 6.2, and 4.1, investigate the seasonal dynamics of the coastal upwelling region along the coast of West Africa in the North Atlantic Ocean. Calculate the (pseudo-) wind stress curl field and compare the winter versus summer wind stress curl field in the vicinity of West Africa. (a) Describe the general character of this seasonal change of wind stress curl along the coast. (b) Make a prediction of what you will see as seasonal changes in SST and chlorophyll, invoking any dynamics you see fit. Next, compare winter versus summer SST and chlorophyll concentration in the same region. (c) Describe the general character of the accompanying seasonal changes in SST and chlorophyll concentration along the coast. How did your prediction fare?

DIVE INTO THE DATA BOX 6.1

Name   MODIS Sea Surface Temperature (SST)

Synopsis High-resolution satellite observations of SST from two satellites carrying the MODIS instrument: Terra and Aqua. The MODIS instrument measures radiance at various wavelengths from the sea surface, in both the shortwave and longwave parts of the electromagnetic spectrum, which is converted to brightness temperature. Empirical algorithms are then used to convert brightness temperature to SST.

DIVE INTO THE DATA BOX 6.1 (cont.)

Science   Gridded data sets of satellite-observed SST are used in a wide variety of scientific investigations, such as understanding the spatiotemporal variability of SST from seasonal to decadal time scales, and at regional to global scales. At such high resolution as that provided by MODIS, relatively fine-scale features (e.g., coastal regions, islands, mesoscale eddies) at the ocean surface field can be studied in detail.

Figure   6.3
Version   Level 3
Variable Sea surface temperature
Platform Satellite
Spatial   Global (gridded), 4 km resolution*
Temporal 2002 to present,** monthly averages***
Source   https://oceancolor.gsfc.nasa.gov/l3
Format   NetCDF (.nc)
Resource modis.m; modis.mat (sample data and code provided on publisher website here: www.cambridge.org/karnauskas)

Journal Reference for Data

Kilpatrick, K. A., Podestá, G., Walsh, S., *et al.* A decade of sea surface temperature from MODIS. *Remote Sens. Environ.* **165**, 27–41 (2015). DOI: 10.1016/j.rse.2015.04.023.

Sample Journal Article Using Data

Jacox, M. G., Moore, A. M., Edwards, C. A., and Fiechter, J. Spatially resolved upwelling in the California Current System and its connections to climate variability. *Geophys. Res. Lett.* **41**, 3189–3196 (2014). DOI: 10.1002/2014GL059589

* 9 km resolution also available.
** Terra MODIS SST record began in February 2000.
*** Daily, weekly, seasonal, and annual averages also available.

DIVE INTO THE DATA BOX 6.2

Name   MODIS Surface Chlorophyll Concentration
Synopsis High-resolution satellite observations of surface chlorophyll concentration from two satellites carrying the MODIS instrument: Terra and Aqua. The MODIS instrument measures solar reflectance

DIVE INTO THE DATA BOX 6.2 (cont.)

at different wavelengths (colors) from the sea surface, and empirical algorithms are used to convert those reflectance values to near-surface chlorophyll concentration.

Science   The chlorophyll concentration near the surface is a proxy for biological productivity. Satellite chlorophyll data sets like that from MODIS therefore enable investigation of the variability of the marine biological environment at local (e.g., coastal regions, islands, mesoscale eddies) to global scales. One is able to investigate how marine productivity responds to physical forcing by combining the data set with other variables such as wind, sea surface height, and SST.

Figures   6.3, 7.3

Version   Level 3

Variable   Surface chlorophyll-a concentration

Platform   Satellite

Spatial   Global (gridded), 4 km resolution*

Temporal   2002 to present,** monthly averages***

Source   https://oceancolor.gsfc.nasa.gov/l3

Format   NetCDF (.nc)

Resource   modis.m; modis.mat (sample data and code provided on publisher website here: www.cambridge.org/karnauskas)

Journal Reference for Data

Hu, C., Lee, Z., and Franz, B. Chlorophyll *a* algorithms for oligotrophic oceans: a novel approach based on three-band reflectance difference. *J. Geophys. Res.* **117**, C01011 (2012). DOI: 10.1029/2011JC007395.

Sample Journal Article Using Data

Del Castillo C. E., Signorini, S. R., Karaköylü, E. M., and Rivero-Calle, S. Is the Southern Ocean getting greener? *Geophys. Res. Lett.* **46** (2019). DOI: 10.1029/2019GL083163.

* 9 km resolution also available.
** Terra MODIS SST record began in February 2000.
*** Daily, weekly, seasonal, and annual averages also available.

7 Coupled Climate Variability

"There's one ocean. There's one atmosphere. They talk."

Susan K. Avery

7.1 Does Natural Climate Variability Matter?

One of the great scientific challenges of our time is understanding our impact on Earth's climate. You might even say it is also one of the great moral imperatives of our time to *do* something with that knowledge. But quantifying the human footprint on climate isn't so simple. Even the most basic number so often used to characterize global warming, the record of global average surface temperature, contains both natural and anthropogenic signals. Put simply, the **anthropogenic** part of any climate record is that attributable to human activities, while the **natural variability** is everything else. If we examine the record of annual mean, globally averaged surface temperature from 1880 to 2017 (Figure 7.1), the upward trend is undeniable, stronger over land than ocean, and quite unrelenting since about 1970. However, the long-term trend since the beginning of the Industrial Revolution is not a straight line. There are slow undulations over decades, and year-to-year variations in global temperature of ~0.5 °C. This superposition of natural and anthropogenic climate signals even in a globally averaged quantity such as this represents one of the roadblocks to identifying, unambiguously, the human footprint on climate. To better understand the sources of natural climate variability is paramount to addressing anthropogenic climate change.

It is the aim of this chapter to use our knowledge of the mechanisms by which the ocean and atmosphere influence one another, described in previous chapters, to dig into some of these flavors of natural climate variability. Beyond the grand challenge of climate *change*, there are also immediate, practical reasons that people and governments are interested in natural climate *variability*. The swings of regional climate at seasonal to decadal time scales have significant impacts on society, ranging from human health (e.g., vector-borne disease transmission), fish stocks, snowpack, freshwater availability (including too much or too little), hurricane activity, and more. Many of the impacts spread deeply into the biogeochemical realm, both on land and sea. For example, interannual climate variability in the equatorial Pacific region strongly modulates marine photosynthesis there, and the

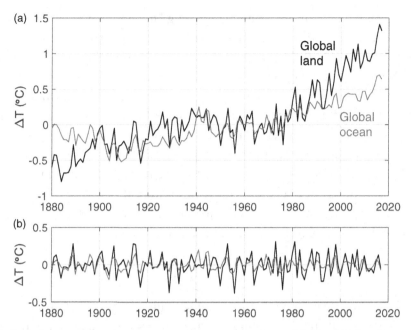

Figure 7.1 Annual mean surface air temperature changes averaged over global land (red) and ocean (blue) areas from 1880 to 2017. (a) The full record; (b) the record with the five-year smoothed records removed, to highlight only the year-to-year variations without the long-term trend. Data from the NASA GISS Surface Temperature Analysis (GISTEMP). All temperatures relative to 1951–1980 base period. (A black and white version of this figure will appear in some formats. For the color version, please refer to the plate section.)

overall exchange of carbon dioxide (CO_2) between the ocean and atmosphere. Most often, such swings are neither like a perfect pendulum nor are they purely random. We like to think there is some predictability in all of this, and of course understanding the physical processes involved is absolutely key to realizing the benefits, like ability to prepare. For many flavors of natural climate variability, the ocean plays a central role. Even when the instrumental record of Earth's surface temperature is separated into global land and ocean realms, we can see clear correlations down to the year-to-year fluctuations. What happens over the ocean is definitely felt on land and, as we will see, the ocean's far-reaching impact on climate is greatly amplified by its *coupling* with the atmosphere. We will touch on many flavors of natural climate variability in this chapter, with a great emphasis on one in the tropics known as ENSO as it encapsulates so completely the essence of coupled ocean–atmosphere variability while serving as an example with wide-ranging effects.

7.2   A Null Hypothesis

Let's start with a null hypothesis about sources of natural climate variability: All climate variability is driven by random "noise." In other words, there are no interesting or consistent physical mechanisms *per se*. One way for this to play out, as described in a set of important papers in the 1970s by German climate scientist

Klaus Hasselmann, is for the ocean and its long "memory" to integrate fast, chaotic weather fluctuations into slow climate variations (Hasselmann, 1976; Frankignoul and Hasselmann, 1977). The source of the ocean's long memory is its enormous heat capacity or thermal inertia – the ocean mixed layer heat budget might be given a *nudge* by a cold snap (cold, dry, and windy weather) lasting a few days, which starts to influence the ocean temperature through surface fluxes encapsulated by Q_{net}, but the weather will move on to the next chaotic fluctuation long before the ocean actually equilibrates with that temporary atmospheric condition. With the ocean's temperature constantly being nudged in random directions by the weather without ever coming into equilibrium, it can stray quite some ways from its average value, and for such periods of time that it even appears to be cyclical. This process can be modeled as follows:

$$\frac{\partial T}{\partial t} = \frac{N - \lambda T}{\rho C_p h} \tag{7.1}$$

where T is the ocean temperature anomaly (deviation away from its average value), N is a small random number that changes each time step t representing those quick punches from the atmosphere, λ is a damping parameter, and $\rho C_p h$ are density, specific heat, and mixed layer depth, respectively – just as phrased in the denominator of our ocean mixed layer heat budget of Chapter 2. From the form of this model equation, we can identify some of the key moving parts. The greater the heat capacity of the water, including its volume ($\rho C_p h$), the less the temperature will be able to change from one time to the next ($\partial T/\partial t$), and the greater the *present* departure of temperature from the average value, the more strongly it will be damped back toward the average value (λT).

Conceptually, this "Hasselmann" model is like flipping a coin many times, but keeping a running tally of the *sum* as you go. Let's do just that – say we have a coin and we'll call heads +1 and tails −1. After 10 flips, the average should start converging on zero, while the cumulative sum (i.e., the sum of all flips, +1 or −1, up to a given point) might have strayed to plus or minus a few (Table 7.1).

After 100 flips, the average will be almost exactly zero, but the cumulative sum might have already strayed to 10 or 20 away from zero. Think of each flip as a random weather anomaly: The cumulative sum is slower to respond to the flips; it will resist straying far from zero unless a fluke of a few heads or a few tails happens

Table 7.1 Results of a coin-flip experiment, showing the result, running mean, and running sum after each of 10 coin flips.

Flip	1	2	3	4	5	6	7	8	9	10
Result	+1	−1	+1	+1	−1	−1	−1	−1	−1	+1
Mean	1.00	0.00	0.33	0.50	0.20	0.00	−0.14	−0.25	−0.33	−0.20
Sum	1	0	1	2	1	0	−1	−2	−3	−2

to clump together – that's the ocean's memory. If we keep this up until we reach one million coin flips (Figure 7.2), and then look back on how that cumulative sum evolved over time, we see that all of those random nudges can indeed move the actual state of the system *well* away from zero and back again, multiple times (without moving the average far from zero, of course).

The Hasselmann model for stochastic forcing of low-frequency climate variability is one possible source of low-frequency (long period) natural climate variability, but there is an important fingerprint of this type of randomly generated variability that distinguishes it from other, more familiar ones that are governed by coupled dynamics. There is no consistent periodicity, and so it is inherently unpredictable. Such a process only serves to "redden" the spectrum of variability,[1] meaning to emphasize the low-frequency (i.e., slow, or long period) variability, but without any robust periods such as a 10-year cycle, or even a blurry one ranging from, say, two to seven years. Since we *do* observe, in our myriad climate records gathered to date, cyclical (and quasi-cyclical) phenomena with clearly identifiable periods, we know there must also be physical mechanisms involved in driving natural variability. In recent decades, we have discovered that ocean–atmosphere coupling is quite often at the heart of those mechanisms. So, we easily reject this null hypothesis, but not without acknowledging its value as a cautionary tale that not every wiggle in a record of natural climate variability can be explained through determinism.

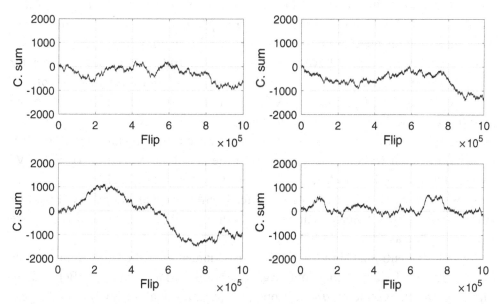

Figure 7.2 Cumulative sum of the results of four sets of one million coin flips. Heads = +1, tails = –1.

[1] Think of a spectrum of climate variability like the spectrum of visible light. The color red is on the low-frequency (or long wavelength) end of the spectrum, so we call a climate signal "red" if it is dominated by a blurring of many slow cycles without the appearance of a single, exact period. In contrast, we refer to a climate signal as having a "white" spectrum if all periods are present (like white light, which is made up of all colors simultaneously).

7.3 Physical Modes of Climate Variability

7.3.1 An Alphabet Soup

Like all self-respecting scientific disciplines, climate scientists love acronyms. Most of the previously alluded to "flavors" of natural climate variability have been given such acronyms. However, unlike the naming of a newly discovered species, drug, or exo-planet, there is no formal set of rules for doing so; in fact, sometimes they are conglomerations of historical terms, inconsistent or arbitrary phrasing, and sometimes there is even more than one acronym for the same climatic phenomenon. So, a little introduction is warranted.

Speaking in a formal statistical sense, we often refer to a quasi-cyclical climate phenomenon as a **mode of variability**. The word mode, in this case, implies some recurring variability in the climate system that has both a spatial and a temporal aspect – a pattern (as in, on a map) that fades in and out over time between two *phases*, like a cycle with a period and an amplitude. As a simple example, think of the basic seasonality of anything on Earth like surface temperature. In January, the Southern Hemisphere is warm and the Northern Hemisphere is cold, and the opposite is the case around July. This is a very regular cycle with a period of one year, and it has a spatial pattern that gradually migrates back and forth between the two opposite phases. This annual mode of variability also has a physical mechanism understood since perhaps the days of Copernicus (the Earth's axis is tilted relative to its orbit around a distant Sun, and perpendicular rays of sunlight are more intense than oblique ones). When enough influential climate scientists are convinced that the mode behaves like an *oscillation* (truly a cycle between exactly two phases), then the acronym ends with "O" for oscillation. One example of this is the North Atlantic Oscillation (NAO), which is a seesaw-like fluctuation in the pattern of atmospheric pressure over the North Atlantic sector. If there are some arguments over whether one is a regular-enough cycle with a plausible mechanism to sustain it, then we might give (or revise) the acronym, replacing "O" with "V" for variability. The Pacific Decadal Oscillation (PDO) is one such example, since it is also frequently referred to as the safer Pacific Decadal Variability (PDV). Of course, some acronyms have simply stuck, and we still use them even if they are not as accurate as they could be.

The modes mentioned above were obviously named after the geographic regions in which their dynamics play out (e.g., the North Atlantic and Pacific) and, in one case, the time scale at which it operates (e.g., decadal, or having a ~10-year period). Sometimes, modes are named after the scientist(s) who discovered them, such as the Madden–Julian Oscillation (MJO), which is an intraseasonal oscillation (i.e., a period of weeks) involving eastward-propagating centers of deep tropical convection over the Indian and Pacific Oceans. Understandably, there are very few examples in the climate science literature pointing to oscillations beyond multidecadal; it is only natural that we would like to observe a good handful of complete cycles before calling some observed swings an oscillation or even a mode of variability. (It is difficult to distinguish a half-cycle from a trend.) Perhaps the more mysterious acronym in the alphabet soup of modes of natural climate variability, in that it contains so little relevant information, is the one defining the most ubiquitous mode of interannual (i.e., multiple years) climate variability on Earth: El Niño–Southern Oscillation (ENSO). We will unpack ENSO in the following subsection.

7.3.2   The Irregular Heartbeat of the Tropics

Before discussing the posterchild for coupled natural climate variability, it is worth briefly reminding ourselves of a few reasons why ocean–atmosphere coupling is likely to be so important in the tropics in the first place. The sea surface is generally warm, so it is easy to get the lower atmosphere to react. There is generally a thermocline within reach, so big changes in sea surface temperature (SST) are possible without even needing to invoke surface fluxes. The atmospheric general circulation moves much more slowly there than in the midlatitudes (which are dominated by strong westerlies and unrelenting storm tracks), so the oceanic budgets have a chance to synchronize – to keep up with the pace of changing atmospheric conditions. The seasonal cycle is relatively small, so coupled mechanisms that are getting underway are less likely to be shocked out of rhythm every six months. Finally, the Coriolis force is there, and even modest atmospheric pressure gradients induced by SST anomalies can set the wind into motion – feedbacks abound.

To understand ENSO, we will apply much of what we have learned in the previous chapters (almost everything, in fact!), but first we must elaborate a bit on the basic state of the wind-driven circulation on the equator, and its coupling to the overlying atmospheric general circulation. This basic state is what we will then *nudge* with something akin to the patch of westerly winds imagined in one of Chapter 6's thought experiments (leading to Kelvin and Rossby waves), and then follow along the processes leading to a full-blown El Niño event.

The equatorial Pacific Ocean and overlying atmosphere is the stage for ENSO. The part of the atmospheric general circulation that is directly coupled to the ocean on this stage is the Walker cell (review Section 5.4.2 on zonal asymmetric and regional circulations). One can imagine the main impetus for the Walker cell as the broad easterly trade winds on the equator, stemming from the converging and westward-deflecting surface branches of the Hadley cell in either hemisphere. The easterly trades are broad enough to drive a northward Ekman transport on the north side of the equator and southward Ekman transport south of the equator; the resulting Ekman transport divergence ($\partial V_{Ek}/\partial y > 0$) leads to Ekman suction ($w_{Ek} > 0$). Through the vertical advection term in the heat budget, this upwelling sustains a cooling of the sea surface ($-w_{Ek}\, \partial T/\partial z < 0$) that results in a steady-state

SST several degrees cooler than average for the tropics. The remarkable zonal continuity yet thinness of this equatorial strip of upwelling is also clear upon mapping the level of biological productivity in the surface waters, which can be done using estimates of **chlorophyll concentration** observed from satellites (Figure 7.3). The long, thin strip of high chlorophyll concentration along the equator is a result of the wind-driven upwelling transporting nutrients (including carbon) upward toward the sunlit surface zone, where phytoplankton use those nutrients in photosynthesis, fixing carbon, and – just like plants on land – generating the green pigment chlorophyll as a byproduct.

The easterly trades would tend to drive equatorial upwelling clear around the world, but of course the presence of landmasses on the equator prevents this, and even enhances it on the eastern sides of oceans by providing barriers away from which surface water can be dragged and replaced by colder water from below. Hence, the wind-driven equatorial upwelling and resulting **cold tongue** dominate the eastern roughly two-thirds of the equatorial Pacific Ocean, whereas the western equatorial Pacific is home to a very large tropical **warm pool** that even extends through the maze of landmasses of Indonesia and into the eastern Indian Ocean. Quite similar to how the Hadley cell emerges as a thermally direct response to a horizontal energy imbalance (Sections 5.3 and 5.4.1), the warm SST in the western equatorial Pacific fuels tropical deep convection and the cool, stable eastern Pacific promotes atmospheric subsidence. A simple constraint of mass continuity is sufficient to connect the ascending and descending branches of the Walker cell with an

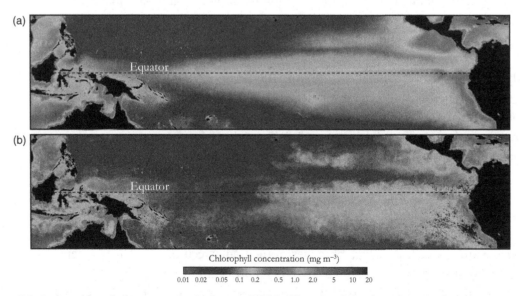

Chlorophyll concentration (mg m^{-3})

0.01 0.02 0.05 0.1 0.2   0.5 1.0 2.0   5   10   20

Figure 7.3 Surface chlorophyll concentration in the equatorial Pacific region during boreal autumn for average conditions (2002–2017 (a)) and during an El Niño event (2015 (b)), from the NASA MODIS instrument aboard the Terra satellite. Non-land areas colored black are due to cloud contamination of the satellite data. (A black and white version of this figure will appear in some formats. For the color version, please refer to the plate section.)

upper branch of eastward flow near the tropopause that gradually descends over the cool eastern equatorial Pacific before rejoining the easterly trades.

The *ocean's* zonal overturning circulation in the equatorial Pacific is almost a mirror image of the Walker cell (Figure 7.4). In addition to the poleward Ekman transports, the easterly trades on the equator drive a westward surface current (the awkwardly named **South Equatorial Current**, or SEC), which actually runs counter to the oceanic zonal pressure gradient set up by the sloping thermocline (shallower in the east due to upwelling). However, that zonal pressure gradient in the ocean must drive an eastward return flow somewhere, and that pressure release happens just beneath the SEC – the **Equatorial Undercurrent** (EUC) flows from west to east and follows the shoaling thermocline as it travels to the Galápagos Islands about 1000 km shy of mainland South America. The EUC also won't meander far from the equator, since any momentary tendency to do so is checked by the Coriolis effect.

So, the Walker cell owes its existence to the zonal SST gradient along the equatorial Pacific, and the zonal SST gradient is likewise sustained through the ocean's dynamic response to the surface branch of the Walker cell. Those are the lead actors on the Pacific stage. They are strongly *coupled* circulations, and so any alteration to part of one circulation is capable of dragging the other into a positive feedback loop. Moreover, the condition of the Walker cell and underlying ocean

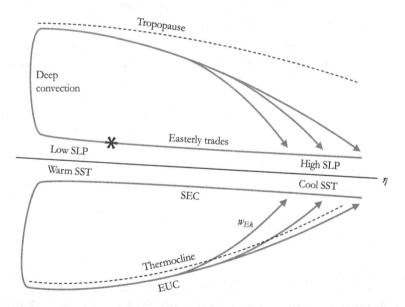

Figure 7.4 Schematic illustration of the zonal overturning circulations of the equatorial Pacific Ocean and atmosphere (i.e., the Walker cell). Acronyms used are SLP for sea level pressure, SST for sea surface temperature, SEC for South Equatorial Current, and EUC for Equatorial Undercurrent. The relative vertical scales between the atmosphere, sea surface slope (η), and ocean are significantly exaggerated; the Walker cell is ~10 km tall while the EUC is only ~100 m deep and η only drops by ~1 m across the Pacific. The asterisk represents the location of a zonal wind anomaly imagined later in this section.

circulation and thermal structure represent the *average* conditions (also known as the baseline, or mean state) of the system. In terms of ENSO variability, this staging would be referred to as **neutral conditions**; there is neither an El Niño nor a La Niña event underway (and we'll define those soon).

Act 1: Born of Kelvin Waves

To find out what El Niño is and how it comes to happen, let's pick up just about where we left off in Section 6.3.2 with the generation of an equatorial Kelvin wave by a westerly wind anomaly symmetric about the equator. In particular, we shall note that this is equivalent to a *slackening* of the easterly trade winds at the location marked by an asterisk in Figure 7.4. We already know what happens next: equatorward Ekman transports converging, Ekman pumping (downwelling), a deeper thermocline ($-w_{Ek}\,\partial T/\partial z > 0$ near the thermocline), and an elevated sea surface mirroring the depressed thermocline, as shown at "Day 1" in Figure 7.5. We have ourselves an equatorially trapped Kelvin wave, which we also know propagates eastward.

Whether our Kelvin wave and its thermocline depression can actually change SST (and evoke a response from the atmosphere) depends on whether there is a vertical gradient in the ocean's temperature (i.e., the thermocline itself) near the surface. In the western equatorial Pacific, a vertical displacement of the thermocline by 10 m or so won't be felt at the surface, since the thermocline is ~100 m deep to start. In the eastern equatorial Pacific, where the thermocline is much shallower and its sloping isotherms even outcrop at the surface, moving the thermocline upward or downward by that amount may cause a change in SST by several

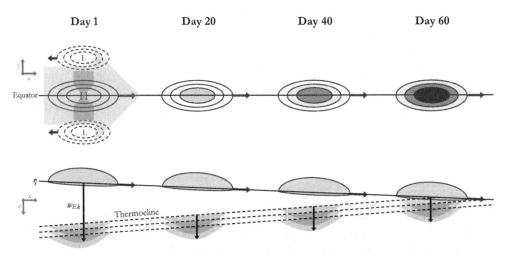

Figure 7.5 Schematic illustration of an equatorially trapped Kelvin wave propagating eastward across the Pacific, warming SST where the mean thermocline is shallow enough. (A black and white version of this figure will appear in some formats. For the color version, please refer to the plate section.)

degrees. So our Kelvin wave, upon reaching the eastern boundary of the equatorial Pacific Ocean, will have produced a modest warm SST anomaly along the latter stretch of its path, covering perhaps the eastern one-quarter to one-third of the basin. In reality, there may be more than one such downwelling Kelvin wave, set into motion by a *burst* of westerly wind anomalies, and because the SST response is dependent on the local vertical stratification $T(z)$, their detection is most reliable in sea surface height measurements such as those made by a satellite altimeter. It has even been suggested that the initial SST warming generated by the first Kelvin wave can tug another westerly wind burst out of the atmosphere, and so on, to produce such sequences of westerly wind anomalies and Kelvin waves rather than just a single, isolated one.

Act 2: Growth by Bjerknes Feedback

Having generated a modest, initial warm SST anomaly in the eastern equatorial Pacific, let's zoom back out and consider what its presence means in terms of the background state (i.e., neutral conditions) of the coupled ocean–atmosphere system as in Figure 7.4. Through suppressed upwelling, we have reduced the coolness (and productivity) of the surface in the eastern equatorial Pacific, which reduces the high-ness of the overlying atmospheric pressure, which reduces the zonal pressure gradient in the atmospheric boundary layer, which reduces the strength of the easterly trade winds. If the easterly trade winds become weaker, then the diverging Ekman transports and resultant upwelling will be weaker, which will *further* warm the cold tongue. In doing so, we have again further reduced the zonal SST and SLP gradients between the warm pool and the cold tongue, which further weakens the Walker cell, and so on. This chain reaction leading to the eventual breakdown of the Walker cell and rise in SST by several degrees in the eastern equatorial Pacific Ocean (Figure 7.6) is known as the **Bjerknes feedback**.

Jacob Bjerknes (1897–1975) described this important, positive feedback in a paper published not terribly long ago (Bjerknes, 1969). In fact, it was in this very paper that he proposed that the episodic surges of warm water along the coast of South America, which had previously been considered a purely oceanographic phenomenon known as **El Niño**, were related to variations in the strength of the Walker cell through coupled ocean–atmosphere interactions! Those variations in the Walker cell were (and occasionally are) considered an atmospheric phenomenon called the **Southern Oscillation**. So the Peruvian fishermen who coined "El Niño" (Spanish for "The Child") and the physical oceanographers to follow were correct in their identification of recurring warm equatorial SST anomalies around Christmastime, and Sir Gilbert Walker (1868–1958) demonstrated impressive acumen by identifying a zonal seesaw pattern in air pressure across the tropics from sparse meteorological observations, but Bjerknes sparked a revolution by putting them together into what we now call the **El Niño–Southern Oscillation**. In this

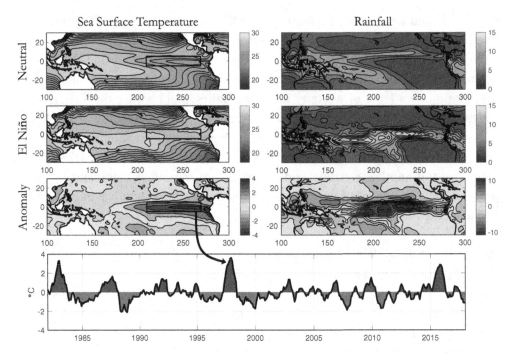

Figure 7.6 Maps of (left) SST (°C) and (right) rainfall (mm/day) for boreal winter neutral conditions (1982–2018 average), El Niño (December 1997–February 1998), and the anomaly (difference between El Niño and neutral). The bottom panel shows a time series of SST anomaly in the NIÑO3 region (i.e., averaged over the eastern equatorial Pacific between 150°W–90°W and 5°S–5°N; box drawn on the SST maps). SST and rainfall observations from the NOAA OIv2 and GPCP products, respectively. (A black and white version of this figure will appear in some formats. For the color version, please refer to the plate section.)

view, we understand that the strength of the easterly trades and the tilt of the equatorial thermocline are dynamically related through zonal momentum and heat budgets, with large anomalies in SST and the Walker circulation appearing every few years as a result. In hindsight, it must have been very difficult to attempt to explain the quasi-cyclical nature, magnitude, or synchronicity of the SST anomalies and wholesale breakdowns of the Walker cell *without* invoking a positive feedback like the one credited to Bjerknes.

Act 3: Death by Rossby Waves
We now find ourselves in the midst of a full-blown El Niño event, with SST in the eastern equatorial Pacific a few degrees above normal and the Walker circulation all but nonexistent – rather, a broad band of rainfall spanning the central Pacific over the now-warmest SSTs, where it is normally cool and dry. We also find a flat, deep thermocline with none of the usual, vigorous upwelling and biological productivity on the equator. How does the coupled system return from such a state to neutral conditions, ready for the next go-round? Yes, there are some negative

feedbacks we can spot in our heat budget framework. For example, the warmer SST emits longwave radiation at a greater intensity, which acts to limit the growth of the SST anomaly. Deep convection and therefore highly reflective clouds reduce the solar radiation cast upon the warmest SSTs. But for the most dramatic of final acts, we recall the pair of Rossby waves that formed off the equator, near the poleward edges of the westerly wind burst(s). It was *from* those areas that the Ekman transports borrowed water in order to construct a Kelvin wave on the equator, so those areas were left with something of an opposite condition – upwelling and low sea surface height anomalies propagating westward. We learned in the previous chapter that Rossby waves propagate westward in order to conserve potential vorticity.

Before proceeding further with the physical process by which Rossby waves can bring El Niño events to a close, let's briefly review the points of contrast between oceanic Kelvin and Rossby waves (Table 7.2). Equatorial Kelvin waves propagate eastward at about 3 m/s, while Rossby waves propagate westward much more slowly – only about 10 percent the speed of Kelvin waves (and slower at higher latitudes). The significance of this tenfold difference in phase speeds is that the pair of Rossby waves does not reach the western boundary of the tropical Pacific Ocean until the Kelvin wave has had ample time to cross the Pacific, warm SST by anomalous vertical advection, and drag the atmosphere into a Bjerknes feedback.

Upon reaching the western boundary, the two Rossby waves turn equatorward and emerge (i.e., reflect) as another equatorial Kelvin wave with – and this is key – the same characteristics they had as Rossby waves in terms of sea surface height anomaly and vertical displacement of the thermocline, which is to say low η and upwelling. They will propagate eastward at the faster pace, as equatorial Kelvin waves must, but they will have the *opposite* impact on the thermocline and SST as the first round of Kelvin wave(s) that generated the El Niño event. The new, upwelling Kelvin wave is thus able to break up the anomalously warm SST/high sea surface height/deep thermocline present in the eastern equatorial Pacific, restore the zonal SST and SLP gradients, reinvigorate the Walker cell, and

Table 7.2 Summary of typical characteristics of oceanic equatorial Kelvin and Rossby waves, including their relative phase speeds, direction of propagation, and location relative to the equator.

	Kelvin	Rossby
Speed	Fast (~3 m/s)	Slow (~0.3 m/s at 10°N)
Direction	Eastward	Westward
Location	On equator	Off equator

restore the equatorial upwelling, thereby maintaining the cold tongue. In many cases observed in the real world, the upwelling Kelvin wave has actually pulled the state of the system *more than* back to neutral – toward the opposite phase of the ENSO spectrum (La Niña) by cooling SST more than normal, at which point the Bjerknes feedback can run in the opposite direction (strengthen the easterlies, tilt the thermocline more, cool SST further, and so on).

The inference, based on our longest instrumental records (~150 years) of SST and SLP variability in the equatorial Pacific, that the preferred period of ENSO is somewhere between two and seven years can be understood, at least in part, by the aforementioned two types of tropical waves and their phase speeds in the ocean. Based on approximate speeds and a distance to travel of ~14,000 km (Papua New Guinea to the Galápagos Islands), it should take an equatorially trapped Kelvin wave about 54 days to cross the Pacific and a Rossby wave at 10°N about another 1.5 years to return. Therefore, it is highly unlikely for a complete ENSO "cycle" (i.e., from El Niño through neutral to La Niña, and back) to be accomplished in a single year. Furthermore – and the scientific community certainly has not settled all of the details on this – there is strong evidence that Earth's annual cycle itself is important in driving or at least modulating ENSO evolution; it is interesting that almost every El Niño and La Niña event tends to peak around December or January! The irregularity in the forcing, such as zonal wind anomalies in the western equatorial Pacific, the asynchronous lags built into the coupled system through differential phase speeds relative to an ocean basin of a particular width, and an apparent once-yearly window of opportunity lends circumstance to the observed 2–7-year "interannual" period dominating historical records of ENSO variability. One thing is for sure: Any suggestion that ENSO is a mere sloshing back and forth of warm water in the tropical Pacific is woefully underestimating the beauty and rich detail of the physical processes at play. After all, waves (including Kelvin waves) do not involve the transfer of mass (only energy), and the SST anomalies are largely the result of local modulations in the vigor of vertical transports *by* those propagating waves and subsequent coupling with large-scale winds.

It is no wonder ENSO is so tempting yet elusive a target to predict, and no wonder so many climate scientists from both the oceanic and atmospheric sciences are still studying ENSO's fundamental dynamics as though it was discovered yesterday! It has a rich set of dynamics, exerts major control on marine biology locally and carbon cycling globally, and is well positioned to generate large seasonal climate anomalies in faraway locations – known as **teleconnections**. Even the intensity and structure of the Hadley circulation is strongly influenced by ENSO (see Figure 5.6 and Section 5.5). It is also important to understand that the above theatrical performance was predicated on a westerly wind burst playing the role of antagonist, but given the nature of a pair of closely coupled circulations like those residing in the equatorial Pacific Ocean and atmosphere, an El Niño or La

Niña event could just as well transpire following other perturbations. Anomalous deep convection in the adjacent tropical Indian Ocean might, for example, stretch the easterly trades out enough to expand the upwelling and cool SSTs in the central equatorial Pacific. Perhaps one of the subtropical gyres to the north or south of the tropical Pacific might carry an anomalously warm blob of water from the subtropics into the eastern equatorial Pacific. It is also important to recognize that every El Niño (and La Niña) event is different, not only in the details of its genesis, but where along the equator the warming is most pronounced and how the global atmosphere adjusts to the tropical heating. No matter where it begins or just what it looks like fully mature, there is always the potential for ocean–atmosphere coupling to take hold in the tropical Pacific, leading to climate anomalies rippling around the planet through the mechanisms for atmospheric responses to SST anomalies, discussed in Section 5.5.

Despite every ENSO event being a bit different from the last, there is at least a *canonical* teleconnection pattern that we can glean from observational data. The **local** atmospheric response *within* the tropical Pacific to the El Niño SST anomalies may be a blend of mechanisms discussed in Section 5.5, including the Gill response and perhaps the SST gradient response (particularly near the surface). More broadly, though, we can see that the entire depth of the troposphere is reacting to the equatorial heating anomaly, well into the higher latitudes, just as predicted by the Gill response and attendant stationary Rossby waves (Figure 7.7).[2] This is what we refer to as the **remote** response. Also notice that the pattern of alternating highs and lows, which bends eastward in the midlatitude westerly flow along their poleward journey, is roughly symmetrical by hemisphere save for the anticyclone positioned off Antarctica being stronger than its counterpart over western Canada. These Rossby waves and their coincidental locations and orientations relative to continental coastlines have very important consequences for seasonal climate anomalies felt at the surface and other features of the atmospheric circulation that people pay attention to, such as the jet stream. Notice, for example, how the westerly flow is accelerated in the corridor between the strong low anomaly in the North Pacific and the anticyclonic anomaly over the tropics to the south. This acceleration is roughly in geostrophic balance and contributed by *both* the cyclonic circulation anomaly to the north and the anticyclonic circulation anomaly to the south. Hence, we *expect* the Pacific midlatitude jet stream

---

[2] Figure 7.7a is a field regression map. Think of it as a map where a value indicated on the map (whether by color or vector) is the slope of the least-squares linear regression line drawn through a scatter plot of the time series of the variable at that location versus a common time series – in this case the NINO3.4 index. So, the color at any location on this map represents the sensitivity of 300 mb geopotential height anomalies there to SST anomalies in the east-central equatorial Pacific Ocean. You can practice calculating field regression maps yourself by tackling Question 15 at the end of this chapter!

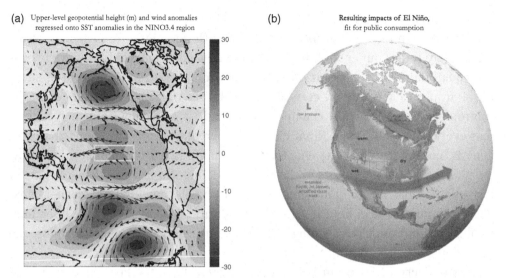

(a) Upper-level geopotential height (m) and wind anomalies regressed onto SST anomalies in the NINO3.4 region

(b) Resulting impacts of El Niño, fit for public consumption

Figure 7.7 (a) Demonstration of the dynamical response of the troposphere to El Niño SST forcing. Field regression of 300 mb geopotential height and wind anomalies onto the time series of SST anomalies in the NIÑO3.4 region (170°W–120°W and 5°S–5°N; box drawn on map). Units for 300 mb geopotential height are meters per standard deviation of variability in the NIÑO3.4 index; positive values indicate anomalous highs or anticyclones and negative values indicate anomalous lows or cyclones. (b) Schematic meant to communicate the likely impacts of El Niño to the general public on the boreal wintertime jet stream and North American surface climate (NOAA Climate.gov). (A black and white version of this figure will appear in some formats. For the color version, please refer to the plate section.)

to be stronger and more elongated during El Niño winters. The warm surface temperature anomaly anticipated for the Pacific Northwest is explained by the same feature: Anomalous northward (geostrophic) flow along the west coast of North America advects warmer, humid air from the south into a region that is normally colder and drier that time of year. The strong anticyclonic circulation positioned over the Amundsen Sea in the Southern Ocean has been shown by a team at the Woods Hole Oceanographic Institution led by Alison Criscitiello to influence the transport of marine aerosols and sea salt onto the West Antarctic ice sheet. In particular, the circulation anomaly is situated conveniently to affect how strongly the winds are blowing onshore and offshore at different points along the Antarctic coastline, favoring the opening of polynyas (Criscitiello *et al.*, 2014). All of the above examples are canonical features of the **remote** response of the climate via global atmospheric teleconnection to ENSO forcing in the tropical Pacific.

Further evidence of ENSO's coupled Earth system dynamics and their global repercussions is given by the biological (and biogeochemical) response to them. In fact, ENSO events are detectable even in the record of *global* atmospheric CO_2 concentration, but for reasons that may be surprising. The eastern equatorial Pacific Ocean is the largest oceanic source of CO_2 to the atmosphere, primarily due to the upwelling of CO_2-rich water increasing the gradient of CO_2 across the air–sea interface (CO_2 flux is proportional to the gradient of partial pressure of CO_2, or pCO_2, between the two fluids; it will flow from higher toward lower partial pressures). During an ENSO event of either phase, there are several physical, chemical, and biological mechanisms by which the ocean-to-atmosphere flux of CO_2 may be altered (Table 7.3). The first is relatively simple: CO_2 is less soluble in warmer water, which would increase the CO_2 flux to the atmosphere during El Niño. Second, the reduced photosynthesis in the surface waters during El Niño (see maps of surface chlorophyll concentrations in Figure 7.3) means leaving behind a higher CO_2 concentration in the surface ocean since it is not being consumed by phytoplankton, which would also increase the CO_2 flux to the atmosphere. However, the reduction of upwelling reduces the transport of CO_2-rich water toward the surface (which is why it is a CO_2 *source* in the first place), which would reduce the CO_2 flux to the atmosphere. Recent satellite measurements show that the above three "direct" processes (i.e., changes to concentration of CO_2 at the surface of the equatorial Pacific Ocean) amount to a net *reduction* in atmospheric CO_2 concentration over the tropical Pacific, so the "less upwelling of CO_2-rich water" is apparently quite important (Chatterjee *et al.*, 2017; Liu *et al.*, 2017).

There is, however, an additional mechanism by which ENSO factors into the *global* atmospheric CO_2 concentration, and the latest science suggests that it dominates in the long run. An El Niño event shifts the distribution of deep tropical convection and rainfall; the ascending branch of the Walker cell follows the warmest SSTs toward the central and eastern equatorial Pacific. This leads to a significant aridification of tropical landmasses in the western Pacific, such as Indonesia, which makes them susceptible to intense droughts and wildfires. Fires consume oxygen, and drought of course means fewer photosynthesizing plants – both of which lead to higher CO_2 concentrations in the atmosphere.

It is possible that, over time, the impact of natural ENSO variability on the global carbon cycle is a wash, given the alternating El Niño and La Niña phases, but it should be taken as another cautionary tale that as the global climate system

Table 7.3 Summary of the different processes by which El Niño impacts the global atmospheric carbon dioxide (CO_2) concentration, including whether it leads to more or less atmospheric CO_2.

Change during El Niño	Impact on atmospheric CO_2 concentration
CO_2 less soluble in warmer water	More
Less biological productivity	More
Less upwelling of CO_2-rich water	Less
Droughts and wildfires	More

evolves *because* of changing atmospheric CO_2 concentrations, there are possible feedbacks *on* the atmospheric CO_2 concentration. Changes in the ocean circulation are just one such pathway, and ENSO serves as a natural laboratory in which we might preview its details.

7.3.3   Other Modes of Tropical Ocean–Atmosphere Variability

There are important modes of interannual climate variability in the other two tropical ocean basins as well, and oftentimes our understanding (and even naming) of them has been colored by how we think about ENSO. In the equatorial Atlantic, this is known as **Atlantic Niño**, and the dynamics very closely resemble those of ENSO. **Benguela Niño** is the name of a localized warming along the African coast ~15° south of the equator, and probably results from warm anomalies sneaking undetected along the thermocline and only being seen at the surface where they finally outcrop in a localized region along the Namibian and Angolan coastlines (Florenchie *et al.*, 2003). In the Indian Ocean, the **Indian Ocean Dipole** (IOD) mode – also known as Indian Niño – has gained a great deal of attention in the last decade or so because of its impact on the monsoon of nearby India and even possible interaction with ENSO. Like ENSO in the Pacific, the alternating phases of the IOD also change which side of the Indian Ocean is rainy and drought-stricken, and where the marine ecosystems are productive or stifled.

What ENSO and its contemporaries in the adjacent tropical ocean basins have in common is a decidedly *zonal* format. ENSO, the other Niños, and the IOD all involve east–west reorganizations of heat, circulation, and rainfall. The 1990s saw the emergence of research on the *meridionally* oriented modes of coupled ocean–atmosphere variability in the tropics, which are referred to as **meridional modes** (Chiang and Vimont, 2004). Meridional modes involve the interplay between the meridional SST gradient near the equator (i.e., between the cold tongue and warmer tropical waters immediately to the north), the trade winds, and the ITCZ (or, equivalently, the rising branch of the Hadley cell). One of the mechanisms for the atmospheric response to SST anomalies discussed in Chapter 5, the horizontal SST/SLP gradient mechanism (step E in Section 5.5), has been invoked to explain this dynamic. When the meridional SST gradient is stronger than normal, the cross-equatorial flow of the trade winds will be stronger than normal (as driven by the stronger-than-normal meridional SLP gradient), and the ITCZ will be shifted a few degrees latitude northward of its normal position. The opposite scenario leads to a southward-shifted ITCZ.

Just a couple of years prior to the first investigations into meridional modes, a new coupled ocean–atmosphere feedback was discovered that proves important in understanding meridional modes and other asymmetries defining the climate of the deep tropics. This is known as the Wind–Evaporation–SST (WES) feedback (Xie and Philander, 1994), and it works as follows (Figure 7.8). Begin with an

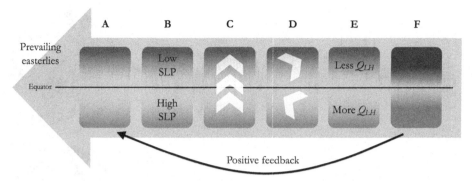

Figure 7.8 Schematic illustration of the Wind–Evaporation–SST (WES) feedback, a positive feedback in the tropics that is capable of amplifying an anomalous meridional SST gradient. (A black and white version of this figure will appear in some formats. For the color version, please refer to the plate section.)

anomalous meridional SST gradient, such that SST is anomalously warm north of the equator and cold south of the equator (step A). A south-to-north pressure gradient force is set up in the atmosphere above (step B), which drives the southerly wind anomaly (step C). The Coriolis force deflects the southerly wind such that the prevailing easterly trade winds are enhanced south of the equator and reduced north of the equator (step D), which increases the latent heat flux Q_{LH} over the anomalously cool SSTs and reduces it over the warm SSTs (step E). The result is a further warming of the warm SSTs north of the equator and a further cooling of the cool SSTs south of the equator (step F), which is just more of step A and the process can begin again. The WES feedback is necessary to explain the amplitude of meridional modes playing out in the tropical Pacific and Atlantic.

7.3.4   Beyond the Tropics

Climate variability in the high latitudes of either hemisphere is dominated by **annular modes**, meaning having a ring-shaped pattern (not to be confused with *annual* modes, meaning with a periodicity of one year). Before looking at them, let's sort out the alphabet soup on these to avoid confusion. Both hemispheres have an annular mode; the one centered over the North Pole is called the **Northern Annular Mode** (NAM) and the one centered over the South Pole is called the **Southern Annular Mode** (SAM). However, the more commonly used names for the NAM and SAM are the **Arctic Oscillation** (AO) and **Antarctic Oscillation** (AAO), respectively. That's not bad, but here's where things start to get interesting. The NAM or AO has an older sibling that is perhaps more widely referred to than the others combined, the **North Atlantic Oscillation** (NAO), which we can better appreciate upon looking at their patterns of spatiotemporal variability (Figure 7.9).

Prior to the wealth of global weather and climate observations that we now have the benefit of, the NAO was defined simply as the *difference* in atmospheric

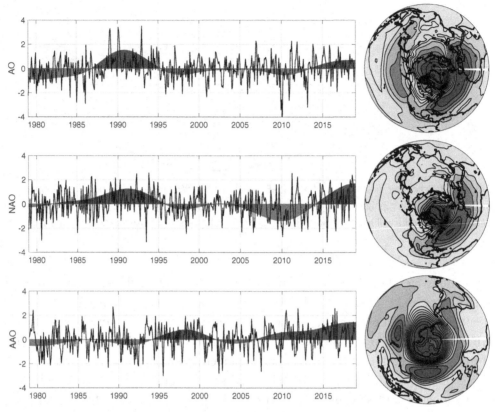

Figure 7.9 Arctic Oscillation (AO, aka Northern Annular Mode [NAM]), North Atlantic Oscillation (NAO), and Antarctic Oscillation (AAO, aka Southern Annular Mode [SAM]). Monthly time series (left column) from 1979 to 2018 with a 10-year smoothing and sea level pressure (SLP) anomaly correlations (right column). The SLP anomaly correlation maps have contour interval 0.1 and the colors saturate at correlation coefficients of ±0.66. Climate indices obtained from the NOAA Climate Prediction Center (CPC), and SLP data from the NOAA/NCEP-DOE Reanalysis 2 data set. (A black and white version of this figure will appear in some formats. For the color version, please refer to the plate section.)

pressures measured in populated cities located beneath the quasi-permanent Azores High and Icelandic Low, such as Lisbon in Portugal and Reykjavik in Iceland. When that pressure difference is large, the NAO is positive, and vice versa. The NAO has chaotic fluctuations across the spectrum from weather (daily) to subseasonal (monthly) time scales (notice how much "noisier" the NAO index is compared to ENSO, yet both have the same monthly temporal resolution). The NAO pre-dates its close cousin, the AO or NAM, for practical reasons. In fact, the history of human curiosity, application, and scientific research on the NAO phenomenon (under various names) goes back many centuries (Stephenson *et al.*, 2003).

Modern observational data with global coverage have since revealed that the NAO has a very large footprint – more than just the weather between Iceland and the Azores. The positive phase of the NAO is characterized by a low-pressure anomaly covering the Arctic Circle and a high-pressure anomaly over the North Atlantic stretching from the interior of North America clear across western Europe. These fluctuations are closely linked to the strength and position of the jet stream, and the kind of weather Europe experiences. We also notice that the NAO appears to have a longer-period aspect of its variability (decadal), but the source of those slower undulations remains an active area of climate research: Do they involve coupling with the ocean? Do they arise from something akin to the Hasselmann model discussed at the beginning of this chapter? Or are they purely statistical artifacts of random noise?

An important aspect of the NAO that has also become clear with the benefit of modern observations and models is that it is likely a regional expression of something even bigger – that is, the NAM or AO. The temporal variability between the NAO and AO climate indices is virtually indistinguishable (both the monthly and decadal fluctuations), and their spatial patterns are also very similar. Due to the way the NAO is defined (i.e., based on pressure differences *within* the North Atlantic region), the pattern is emphasized there while the pattern that emerges from the NAM/AO is broader over the Arctic and more zonally symmetric, including a similar center of action over the North Pacific. One could argue that humans invented the NAO construct out of our temporary ignorance of the annular modes, and if modern civilization unfolded differently, we might well have "discovered" a North *Pacific* Oscillation centuries ago, only to later realize it was just a part of the AO/NAM. Nonetheless, it has stuck and the NAO will probably never fade from the canon of climate dynamics.

The SAM or AAO pattern is essentially the same as its Northern Hemisphere counterpart, but apparently operates independently since the AO and AAO indices are perfectly uncorrelated. The current wisdom on both of the annular modes (and the NAO) is that they are *not* fundamentally driven by ocean–atmosphere coupling, but they definitely have significant impacts on the ocean by way of wind stress and surface fluxes of both heat and freshwater, with a large yet (for now) less well-defined potential for feedbacks.

There are several flavors of natural, decadal (and multidecadal) climate variability that are the subject of intense research and discussion, and these involve physical mechanisms spanning from the tropics to the higher latitudes. In all cases, ocean–atmosphere coupling is suspected to play an important role. Decadal variability in the Pacific has been coined the **Pacific Decadal Oscillation** (PDO), or Pacific Decadal Variability (PDV) if one prefers not to imply too much, too soon regarding its temporal behavior or physical mechanisms. More recently, the PDO has been augmented with a secondary feature called the **North Pacific Gyre Oscillation** (NPGO), which redefines the decadal variability in such a way as to connect more directly to impacts such as coastal upwelling and associated ecosystem dynamics (Di Lorenzo *et al.*, 2008). The spatial pattern of the PDO in terms of SST anomalies stretches from the equator to the high-latitude North Pacific, so unsurprisingly many have hypothesized that it is linked to ENSO. The linkage between the tropics and midlatitudes in this case might be directly through the ocean, or via the atmospheric response to SST anomalies in one region serving to influence a distant oceanic region – known as an **atmospheric bridge**. However, we would still need a physical mechanism for its slower fluctuations than those of ENSO (i.e., decadal, rather than just interannual).

The **reemergence mechanism** has been proposed as a way for the ocean to effectively "remember" SST anomalies across multiple years (Newman *et al.*, 2016). This widely cited mechanism works as follows. Somewhere in the midlatitudes during wintertime, the atmosphere imparts an SST anomaly on the ocean through some component of the net surface flux Q_{net} (more solar radiation Q_{SW} than normal, for example). Given the windy conditions typical of midlatitude winter, this temperature anomaly is quickly spread out throughout the depth of the mixed layer, which is relatively deep in winter. Transitioning into summer, the mixed layer depth shoals and leaves behind (or below) the temperature anomaly, where it is temporarily hidden from SST and the atmosphere. The next winter, the mixed layer again deepens, and the temperature anomaly is entrained back into the mixed layer and therefore factors into the SST. In this way, otherwise short-term surface flux anomalies are able to influence SST across multiple years. There are many other hypothesized mechanisms for natural, decadal climate variability in the Pacific sector; perhaps the dynamical tools acquired through this textbook will allow you to form one of your own.

Multidecadal climate variability in the Atlantic sector has long garnered the attention of the climate science community. One reason (of many) for this is that any variations in SST directly impact the maximum potential intensity of hurricanes, and the slower the SST variations, the more attractive they are to those trying to make long-range predictions. The marquee event in the Atlantic sector is the **Atlantic Multidecadal Oscillation** (AMO), or just the more cautious Atlantic Multidecadal Variability (AMV). The AMO has a relatively long time scale, perhaps a period of 60–80 years, although we are unable to pin it down much more precisely than that, because our instrumental SST records are only ~150 years long and so we can only directly observe a couple of complete cycles. The calculation of the AMO index is quite simple: Traditionally, it is just SST averaged over the entire North Atlantic Ocean (from the equator all the way to 80°N). The spatial pattern of the AMO is not much more complicated than that, either, although when the AMO is in a positive phase, the *far* North Atlantic Ocean (near the mouth of the Labrador Sea) is where the warmest SST anomalies are found. Like all of the flavors of natural climate variability mentioned thus far, the AMO has wide-ranging global impacts from the climates of North America, Europe, and Africa, to sea ice concentrations in the Arctic.

Much of the AMO remains shrouded in mystery, partly because of its long time scale relative to the human experience and the length of our instrumental data sets. We are reasonably confident that it is independent from the NAO, but we are also very suspicious that it might be closely related to (if not driven entirely by) the "other" component of the global ocean circulation: the density-driven – or thermohaline – circulation. The part of the thermohaline circulation that resides in the Atlantic is called the Atlantic Meridional Overturning Circulation (AMOC). Perhaps it is not a coincidence that there is so much overlap in the acronyms of AMO and AMOC. In the next chapters, we will explore the dynamics of the buoyancy-driven ocean circulation and see how the field of paleoclimate adds insight to our understanding of such climate variations as the AMO.

Further Reading

A special issue of *Science* magazine was devoted to the Orbiting Carbon Observatory-2 (OCO-2), measuring the impact of ENSO on atmospheric CO_2 (Vol. 358, Issue 6360).

Several excellent textbooks are dedicated entirely to the topic of ENSO, including those by Clarke (2008) and Sarachik and Cane (2010).

For a broad survey of advances in understanding of coupled ocean–atmosphere variability, see the monograph *Earth's Climate: The Ocean–Atmosphere Interaction* (Wang *et al.*, 2004).

Questions

1. Explore the potential for random noise to translate into coherent low-frequency variability, based on the "coin-flip" adaptation of the Hasselmann model. (a) How many coin flips (1, 10, 100, etc.) do you find necessary for low-frequency variability to emerge clearly to the naked eye? (b) Very roughly sketch a graph of one of your time series with low-frequency variability. Indicate how many coin flips were used in this case. (c) What is the approximate amplitude of the low-frequency variability for the following numbers of coin flips? Answer for: 10, 100, 1000, 10,000, 100,000, and 1,000,000 flips. (d) Do you see cyclic (periodic) oscillations? At approximately what period(s) do you see cycles? The following sample MATLAB script may be used and modified to suit your needs.

```
flips=round(rand(1000,1));
flips(flips==0)=-1;
figure(1)
subplot(2,1,1); bar(flips); ylim([-1.5 1.5])
subplot(2,1,2); plot(cumsum(flips))
```

2. Oceanic Kelvin waves can be either "upwelling" waves or "downwelling" waves. In simple terms, what phenomenon in the atmosphere determines whether a Kelvin wave has an upwelling (which cools SST and might lead to a La Niña) or downwelling (which warms SST and might lead to an El Niño) anomaly associated with them?

3. Identify two feedbacks involved in ENSO dynamics, one positive and one negative.

4. Sketch and narrate a storyboard (a few diagrams and key points) that explains the dynamics leading to (and bringing to an end) a La Niña event.

5. A recent NASA satellite mission called the Orbiting Carbon Observatory 2 (OCO-2) shed light on CO_2 exchanges during a major El Niño event. Briefly describe *one* of the ways in which ENSO influences atmospheric CO_2.

6. In the real world, ENSO is not a perfectly regular oscillation. What phenomena do you think might contribute to any irregularity (and difficulty predicting) in ENSO?

7. The schematic illustration ("fit for public consumption") shown in Figure 7.7b is obviously biased toward ENSO impacts over North America. Based on the field regression map shown in Figure 7.7a, what kind of air temperature anomaly (warm or cold) would you expect in New Zealand during an El Niño? Explain why.

8. It may be the case that the NAO and AO (or NAM) are slightly different takes on the same phenomenon, but what about the traditional definition of the NAO accounts for its much earlier discovery than the others?

9. In simple terms, what is special about the equator that enables the Wind–Evaporation–SST (WES) feedback to be such an efficient mechanism to amplify meridional (cross-equatorial) SST gradient anomalies there?

10. Select one of the *Dive into the Data* boxes featured in this chapter. Read the sample journal article that uses the associated data set, and describe the general role of that data set in the study. Provide a little context, both technical and scientific. What did the authors actually "do" with the data set? What specific scientific insight was enabled by incorporating this data set into the study?

11. Utilizing *Dive into the Data* Box 3.4, examine the time-mean structure of zonal wind stress τ_x and zonal ocean velocity u along the equator in the Pacific Ocean. (a) Can you spot the Equatorial Undercurrent (EUC) beneath the westward surface current? What is its depth, and how does that depend on longitude? (b) How does the Pacific EUC compare to any analogous circulation features in the equatorial Indian and Atlantic Oceans? (c) Describe the changes in τ_x and the EUC during the peak of an El Niño event (e.g., January 1998), giving both qualitative description and rough numbers to quantify the changes you observe.

12. Following on from the previous question, utilize *Dive into the Data* Boxes 2.3 and 4.3 to complement your analysis of circulation changes during an El Niño event with changes in important dynamical fields like equatorial sea surface height and thermocline depth. (a) How do these quantities along the equatorial Pacific change during the height of the El Niño? (b) In terms of the zonal momentum budget, can you use such changes in equatorial η and thermocline depth to diagnose the circulation changes observed in the previous question?

13. Utilizing *Dive into the Data* Box 2.2, calculate the four common ENSO indices based on SST anomalies: NINO1+2, NINO3, NINO3.4, and NINO4. Explore their variability by making a figure that overlays these four time series. (a) How do the four indices compare to one other in terms of their amplitude of variability (e.g., standard deviation)? (b) Produce a global map of some measure of the variability of SST anomalies. Does this map confirm your results from part (a) of this question? (c) Which two ENSO indices have the lowest temporal correlation with one another? (d) Define a criterion (or set of criteria) for an El Niño event. How many such events occurred from 1982 through 2018, and what period (as in 1/frequency) does that imply for ENSO as an oscillation? Does your simple estimate of ENSO period fall within the widely cited ENSO period of 2–7 years? (e) Use the internet to find the latest ENSO forecast from the NOAA Climate Prediction Center (CPC) or the International Research Institute for Climate and Society (IRI). If there is a forecasted El Niño or La Niña on the horizon, how does it compare to past events? Would it qualify as an El Niño based on your definition in part (d) of this question?

14. Utilizing *Dive into the Data* Boxes 2.2 and 5.4, examine the relationship between zonal SST and SLP gradients in all three major tropical ocean basins. Can a simple yet quantitative empirical model of ocean–atmosphere coupling be derived from these quantities?

15. Utilizing *Dive into the Data* Boxes 2.2 and 5.3, produce a field regression map of OLR anomalies onto the NINO3.4 index. How would you explain the pattern of OLR response that you see in terms of ENSO-related regional climate anomalies and why they might locally alter the amount of longwave radiation escaping to space?

DIVE INTO THE DATA BOX 7.1

Name   NASA GISS Surface Temperature Analysis (GISTEMP)
Synopsis   An important record of Earth's globally averaged surface temperature since the nineteenth century, constructed by blending surface temperature observations over land (weather stations) and the

DIVE INTO THE DATA BOX 7.1 (cont.)

	ocean. NOAA ERSST v.5 is used over the ocean (see Box 8.5). The NASA Goddard Institute for Space Studies (GISS) updates GISTEMP each month.
Science	The GISTEMP record is one of the prominent data sets quantifying global warming. The scale of the warming trend can be compared with natural (year-to-year and decadal) fluctuations, along with the cooling influence of major volcanic eruptions and emission of anthropogenic aerosols.
Figures	7.1, 8.9, 9.8
Version	January 2019*
Variable	Globally averaged surface temperature
Platform	Weather station, satellite, ship, mooring, float
Spatial	1D (time series), global land and ocean**
Temporal	1880 to present, annual averages***
Source	https://data.giss.nasa.gov/gistemp
Format	Plain text (.txt) and comma-separated values (.csv)
Resource	gistemp.m; gistemp.mat (sample data and code provided on publisher website here: www.cambridge.org/karnauskas)

Journal Reference for Data

Hansen, J., Ruedy, R., Sato, M., and Lo, K. Global surface temperature change. *Rev. Geophys*. **48**, RG4004 (2010). DOI: 10.1029/2010RG000345.

Sample Journal Article Using Data

de Jong, M. F. and Steur, L. Strong winter cooling over the Irminger Sea in winter 2014–2015, exceptional deep convection, and the emergence of anomalously low SST. *Geophys. Res. Lett.* **43**, 7106–7113 (2016). DOI: 10.1002/2016GL069596.

* Version as of May 2019 is v.3; v.4 is forthcoming.
** Also available as monthly, global, gridded fields in NetCDF format.
*** Monthly averages also available for the globally averaged (land and ocean combined) surface temperature.

8 | Response to Buoyancy Forcing

"The climate system is an angry beast, and we are poking it with sticks."

Wallace S. Broecker (Stevens, 1998)

8.1 Overview of Buoyancy Forcing

What is buoyancy forcing, and why do you already know quite a bit about it? In Chapter 2, our goal was to understand the physical processes that set the *temperature* of seawater. The task of Chapter 3 was similar, but for *salinity*. In Chapter 4, we applied a similar "budget" framework to the *velocity* of the ocean. The latter budget, the momentum budget, opened the door to two major drivers of ocean circulation: wind stress and buoyancy. The global distribution of wind stress was rationalized in Chapter 5 from the perspective of the atmospheric general circulation, and in Chapter 6 we saw just how it drives a major portion of the ocean circulation. The **buoyancy** of a parcel of water embedded within more water – that is, whether it will rise or sink – is determined by its density compared to that of the water beneath it. Recall from Equation 4.19c that if the density of a parcel of water is great enough to couple with gravity ($-\rho g$) and overcome the upward thrust of the ever-present vertical pressure gradient in the ocean ($\partial p/\partial z$), then it shall sink. As it turns out, the density of seawater is almost exclusively a function of its temperature and salinity, so Chapters 2 and 3 represented a framework for diagnosing the *buoyancy* of seawater as well.

We needn't be so abstract. The real subject of this chapter is the **thermohaline circulation**, which plays an enormous role in shaping global climate and is probably relevant to how it varies over time, especially over relatively long time scales (decades and longer). It is also propelled by buoyancy forcing (think "thermo" for heat and "haline" for salt). In fact, it would be unusual for the thermohaline circulation to *not* be mentioned in a scientific paper or discussion of dramatic changes recorded in distant past climates, or of how the global ocean circulation and climate might evolve over the *coming* centuries as the planet continues to warm due to anthropogenic emissions of greenhouse gases. The global thermohaline circulation will be described later in this chapter, but first let's have a look at the nature of the relationship between seawater density and its constituent factors: temperature and salinity.

In *principle*, we should expect density to be inversely proportional to temperature (colder = denser) and directly proportional to salinity (more saline = denser). In *practice*, density is calculated from the **Equation of State**, which is a nonlinear function of temperature, salinity, and pressure. There are dozens of high-order polynomial terms in the Equation of State, and their coefficients are determined empirically (such as from *in-situ* and laboratory measurements, rather than derived from physical principles), so we won't even bother writing it down. Instead, we can simply plot the solution to the Equation of State in temperature vs. salinity space and examine its general behavior (see the gently curving black lines in Figure 8.1). Across the spectrum of temperature and salinity values seen throughout most of the world ocean, those intuitive principles hold, but the nonlinearities are also evident. Notably, at low temperatures such as those found in polar regions or the deep ocean, density does not change by as much for an incremental change in temperature than it does in warmer regimes. For example, holding salinity constant (say, 34 g/kg), density varies by ~1 kg/m³ at 0–10 °C, but by nearly 3 kg/m³ at 20–30 °C.

The nonlinearity evident in the T–S diagram also gives rise to the process of **cabbeling**, whereby the mixing of two water masses with equal density results in a water mass with greater density. Note that cabbeling can only occur if the two

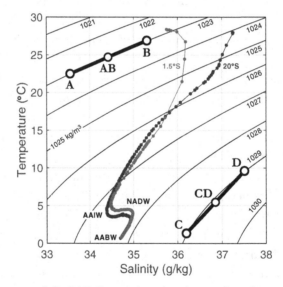

Figure 8.1 Temperature–salinity (T–S) diagram, based on the Equation of State, giving the seawater density ρ (kg/m³) for a wide range of temperatures (0–30°C) and salinities (33–38 g/kg) for pressure at mean sea level (1013.25 hPa). Heavy black features illustrate cabbeling, or the process of two water masses with the same density mixing to create a water mass of greater density. Two hydrographic profiles are shown from the South Atlantic Ocean (WOCE Line A16C, R/V Melville, 13 March to 19 April 1989, Chief Scientist Lynne Talley): one at 25°W, 20°S (blue) and one at 25°W, 1.5°S (red). Temperature and salinity values plotted roughly every 10 m depth. Water masses labeled include North Atlantic Deep Water (NADW), Antarctic Intermediate Water (AAIW), and Antarctic Bottom Water (AABW). (A black and white version of this figure will appear in some formats. For the color version, please refer to the plate section.)

original water masses have the same density *for different reasons* – different combinations of temperature and salinity yielding the same initial value of density. Cabbeling is more effective at increasing the density of seawater through mixing in cold regimes. For example, the mixing of water masses A and B (Figure 8.1) in relatively warm water (~25 °C) produces water mass AB that has virtually the same initial density as A and B (1023 kg/m³). In colder (and denser) water, the mixing of water masses C and D (each with density 1029 kg/m³) produces water mass CD with density greater than 1029 kg/m³. As we will see later in this chapter, sinking of negatively buoyant water is a key driver of the thermohaline circulation, so this represents one way in which mixing can play an important role in global climate, despite substantial challenges involved in measuring or parameterizing it.

8.2   Mathematics of Buoyancy Forcing

. It is something of a mantra of ours to say that "temperature and salinity determine density," and "density is important for the ocean circulation" – especially a component of the ocean circulation that plays a key role in moderating Earth's climate. This is *part* of why developing an intuitive understanding of the ocean mixed layer heat and salt budgets is emphasized in earlier chapters, although sea surface temperature (SST) and sea surface salinity (SSS) are also important variables for other reasons, of course. Now let's put some mathematics to that assertion, so we can quantify how the surface ocean density field is shaped by ocean–atmosphere interactions in the real world, and how that contributes to the thermohaline circulation.

A clever mathematical construct was developed in the 1980s by oceanographers and climate scientists interested in the hydrological cycle, which allowed us to quantitatively relate the ocean–atmosphere fluxes of heat and freshwater directly to those of density. The notion of a "density flux" might sound odd, especially compared to fluxes of energy (heat) and mass (freshwater or salt). A simple analogy may be helpful, using a quantity that is important yet difficult (if not impossible) to measure directly: the distance Δr between a hiker and a bolt of lightning just observed to strike the ground. We already know the relationship between distance traveled Δr and time elapsed Δt before the sound of thunder reaches the hiker (i.e., the speed of sound) from theoretical and empirical studies beginning, as usual, with Isaac Newton. Therefore, all the hiker needs to measure is Δt in order to know the distance Δr from the lightning bolt using the rational equation

$$\Delta r = \frac{\Delta r}{\Delta t} \Delta t \qquad (8.1)$$

where $\Delta r/\Delta t$ is, of course, the speed of sound. What are the uncertainties in such an estimate? Some amount of time elapsed between the actual strike and when the light(ning) reached the hiker's eyes, but light travels nearly a million times faster than sound, so we can neglect that. The nervous hiker might be making a very rough estimate of Δt based on mental counting, introducing a possible measurement error. Finally, the speed of sound is not a universal constant; the speed of such compression waves depends nonlinearly on various parameters of the medium through which it propagates, such as temperature, humidity, and pressure, so a state-dependent value of $\Delta x / \Delta t$ could be used. Such dedication to accuracy would be impressive, but overkill when determining whether to seek shelter from imminent danger – five seconds per mile is good enough!

The above is a surprisingly faithful analog to how density (and buoyancy) fluxes are computed. If salinity had a twin sibling living in the atmosphere, it would be humidity, and from the Equation of State (plotted on Figure 8.1 for standard pressure at sea level), the relationships between density and temperature, and between density and salinity, themselves depends on the concurrent temperature, salinity, and pressure. If we can measure heat and freshwater (or salt) fluxes, and we know the relationship between those fluxes (or their associated state variables, temperature and salinity) and density, then we can quantify the density flux. Fortunately, we now *can* measure surface heat and freshwater fluxes routinely and globally, and the Equation of State yields our required "rate" coefficients, $\Delta \rho / \Delta T$ and $\Delta \rho / \Delta S$. The **density flux** F_ρ is thus given by

$$F_\rho = \alpha F_T + \beta F_s \tag{8.2}$$

where α and β are the thermal expansion and haline contraction coefficients, respectively, and F_T and F_S are the measured heat and salt fluxes, respectively. αF_T thus represents the contribution of surface heat flux to density flux, and βF_S represents the contribution of salt (or freshwater) flux to density flux. The fact that temperature and salinity are sufficiently independent of one another allows us to write density flux as a simple linear combination of the thermal and haline density flux components.

The coefficients α and β are then defined as partial derivatives representing the functional relationships

$$\alpha = \frac{\partial \rho}{\partial T} \tag{8.3}$$

$$\beta = \frac{\partial \rho}{\partial S} \tag{8.4}$$

These coefficients therefore play the role of the speed of sound in the previous analogy; they allow us to convert a given heat flux to a density flux, and a given salt flux to a density flux, and their values are state-dependent just as shown on the T–S diagram (Figure 8.1). Since density decreases as temperature increases, α takes a negative value, while β takes a positive value as density increases with salinity. The heat and salt fluxes F_T and F_S take on a familiar form relative to the mixed layer heat and salinity budgets presented in Chapters 2 and 3, respectively:

$$F_T = \frac{Q_{net}}{\rho C_p} \tag{8.5}$$

$$F_s = S(E - P) \tag{8.6}$$

Assembling the complete expression for density flux including all of the coefficients and fluxes yields

$$F_\rho = \frac{\partial \rho}{\partial T} \frac{Q_{net}}{\rho C_p} + \frac{\partial \rho}{\partial S} S(E - P) \tag{8.7}$$

where the units of density flux F_ρ work out to kg m^{-2} s^{-1}. Adopting the common definition of *flux* – a rate (per seconds) of transfer of some quantity through a unit area (per m^2) – F_ρ is technically a *mass* flux (or a density flux multiplied by one cubic meter), but density flux is the language that has permeated the scientific literature since the late 1980s, when oceanographer and global water cycle expert Ray Schmitt first applied Gösta Walin's trailblazing 1982 derivation to large-scale observations of heat and freshwater fluxes over the North Atlantic Ocean (Walin, 1982).

Ray Schmitt of the Woods Hole Oceanographic Institution, together with colleagues from the Scripps Institution of Oceanography across the continent in La Jolla, California revealed the dominant feature of the density flux field to be a large and persistent density gain ($\sim 14 \times 10^{-6}$ kg m^{-2} s^{-1}) over the Gulf Stream leading toward the Arctic (Figure 8.2). By also mapping out the thermal and haline density fluxes, αF_T and βF_S, separately, Schmitt and colleagues revealed that this feature is overwhelmingly accounted for by thermal density flux, i.e., loss of heat from the surface ocean, lowering its temperature and increasing its density (Schmitt *et al.*, 1989). Given the relatively limited set of heat and freshwater flux observations available at the time, these calculations have been repeated many times over the years, but the general conclusions have stood the test of time and even transcended to aptly characterize the distribution of density flux across the other major ocean basins.

The final touch we can place on the clever density flux construct is to couple it with gravity, and express it as a ratio to a reference density ρ_0 to yield

$$F_B = -g \frac{F_\rho}{\rho_0} \tag{8.8}$$

where F_B is the **buoyancy flux** with units W/kg (i.e., power per unit mass). The sign of buoyancy flux determines whether the ocean–atmosphere fluxes encourage sinking or rising motion in the vicinity of the mixed layer. Vertical flows being driven by buoyancy forcing (independent of fluid medium such as air or water) is called **convection**. If $F_B > 0$, the mixed layer is positively buoyant, and convection is inhibited (indeed further stratification may result). If $F_B < 0$, the surface water is negatively buoyant, and convection may occur. Regardless of sign, a strong vertical stratification (represented as a large reference density ρ_0) would render the right-hand side of Equation 8.8 small, reducing the tendency for convection.

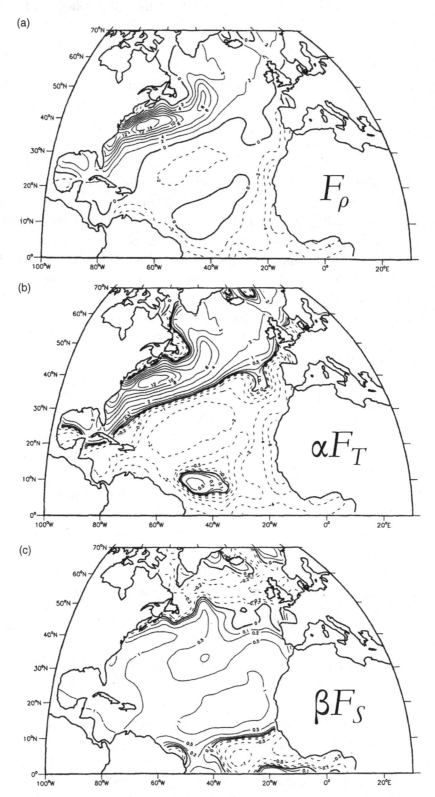

Figure 8.2 Maps of total annual average density flux (kg m^{-2} s^{-1}) (a), thermal density flux (b), and haline density flux (c). Figure adapted from Schmitt *et al.* (1989). © American Meteorological Society. Used with permission.

Physical processes within the ocean's interior can also govern seawater density, such as advection and diffusion, but this buoyancy flux is the mathematical point of contact for the atmosphere to influence the thermohaline circulation; we will now explore its real manifestations on Earth.

8.3 The Global Thermohaline Circulation

There are some interesting – and perhaps useful – parallels between the ocean's thermohaline circulation and the familiar Hadley circulation of the atmosphere. Both are fueled by convection. Recall from Chapter 5 that the atmospheric general circulation is driven, to first order, by the latitudinal gradient of surface heating. Surface air near the equator is warm and moist, and therefore less dense, so it is positively buoyant and *rises* away from the surface (Figure 8.3). In the ocean, convective flows can also be propelled by surface water that is denser and negatively buoyant and therefore *sinks* away from the surface. Each ocean basin is quite different, but on average, the latitudes at which surface seawater has the greatest density (and is likely to sink) are ±60–80°. In the Atlantic sector, the densest surface waters are found along the Antarctic sea ice edge in the Southern Ocean and at two locations in the North Atlantic: near the mouth of the Labrador Sea (just south of Greenland) and in the Nordic Seas (east of Greenland). Regions of "light" surface water (relatively low

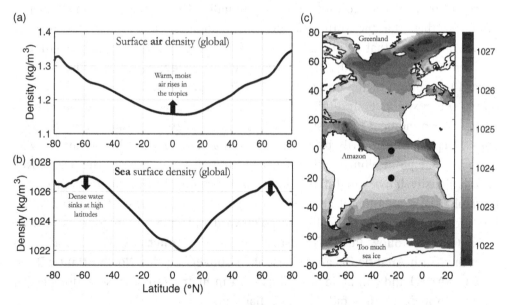

Figure 8.3 Zonal mean profiles of surface air density (a) and sea surface density (b) averaged over the period 2012–2014. (c) A map of sea surface density (kg/m³) in the Atlantic sector for the same period. Sea surface density calculated using the Equation of State based on observed sea surface temperature (NOAA OIv2) and sea surface salinity (Aquarius). The locations of the two hydrographic profiles shown in Figure 8.1 are marked as black circles on the map. (A black and white version of this figure will appear in some formats. For the color version, please refer to the plate section.)

density) are found predominantly in the tropics, where the water is warm and there is abundant input of freshwater – directly beneath the Intertropical Convergence Zone (ITCZ) and near the mouths of major rivers such as the Amazon. If you find it unsettling that all of the aforementioned regions of dense water are positioned next to huge stocks of freshwater (almost 30 million cubic *kilometers* of ice sitting on Greenland and Antarctica), you are certainly not alone – more on this later.

How does surface water in the aforementioned regions become dense enough to sink in the first place? From the previous sections (and Figure 8.1), we know this can be achieved through lowering the temperature, increasing the salinity, and/or mixing, and that one may be more effective than the other depending on geographic context or climatic regime. There are a few common mechanisms by which these can happen in the world ocean that are especially relevant to the thermohaline circulation. The northward-flowing surface water of the subtropical gyres (i.e., the western boundary currents) loses substantial heat through surface turbulent heat fluxes. Cool, dry air flowing eastward off the North American continent enhances both sensible and latent heat fluxes as the warm water travels toward the Arctic (Section 2.2.2; Figures 2.5 and 2.6). This mechanism represents a densification of a surface water parcel by moving it "downward" along the y axis of the T–S diagram (Figure 8.1). At the same time, the evaporation associated with the latent heat flux removes freshwater from the ocean, thereby concentrating the salt left behind in the seawater – a rightward movement on the T–S diagram. The result is water becoming denser the farther it flows poleward toward the seas surrounding Greenland. It should not be too surprising, then, that the North Atlantic Ocean is one of the key convective regions of the world ocean that are thought to propel the overall thermohaline circulation.

When a water mass sinks in this way, we call it **deep water** because it sinks toward the bottom of the ocean while flowing equatorward. The final densification in the mixed layer and initial sinking itself is dubbed **deep water formation**. It may take a moment to digest this unusual vocabulary – we are calling a water mass "deep" not necessarily based on its depth at a given time, but because it will be so at some point along its slow journey through the interior of the ocean. Not only is density important for the downward *propulsion* (i.e., convection) of the thermohaline circulation, it determines the *path* its water masses take through the interior of the ocean. Specifically, the downward and equatorward journey of deep water tends to follow lines of constant density, or **isopycnals**. Water that sinks from the North Atlantic Ocean (as described above) is called **North Atlantic Deep Water** (NADW) and can be identified as such in hydrographic observations based on where the data points fall on a T–S diagram.

Another key part of the world ocean where deep water formation contributes to the thermohaline circulation is the Southern Ocean, consistent with the global distribution of sea surface density (Figure 8.3). Although convection following densification of surface waters is driving this deep water formation as well (like

NADW), other mechanisms are in play due to the unique geography and seasonal sea ice ringing the continent of Antarctica. Sea ice formation is analogous to evaporation: freshwater leaves the liquid ocean, leaving behind a saltier mix of seawater (rightward movement on the T–S diagram), which naturally has higher density. The process of sea ice forming and leaving the salt behind is known as **brine rejection** – chemistry tells us that salt doesn't fit into the crystalline structure of frozen water. At such high latitudes (60–80°S), the water is also very cold, which means increasing salinity via brine rejection is an especially effective way to increase density and encourage sinking. Finally, as cold, dry winds blow off of Antarctica, openings in the sea ice form, which are called **polynyas**. This open water, exposed to the frigid and windy atmosphere, is subject to intense cooling by surface turbulent heat fluxes, which further increases the density and causes negatively buoyant water to sink. The deep water that forms in the Southern Ocean through brine rejection and polynya cooling is called **Antarctic Bottom Water** (AABW); the superlative "bottom" was given to this water mass because it is ever so slightly more dense than, say, NADW, so it literally flows away from Antarctica along the bottom of the ocean.

While in principle, both the physics and the space occupied by the thermohaline circulation are global in nature, one regional subset of the thermohaline circulation tends to get a lot of attention from the ocean and climate research community. The part of the thermohaline circulation that exists in the Atlantic (all the way from the Nordic Seas to the Weddell Sea of Antarctica) is known as the **Atlantic Meridional Overturning Circulation** (AMOC). The AMOC is not independent or detached from the broader thermohaline circulation, which rambles through all of the major ocean basins, but there are some justifiable reasons for studying it as a single entity. For one thing, the Atlantic Ocean is much narrower than the Pacific Ocean, and there are many developed nations situated directly adjacent to it with a longstanding interest in its circulation (not just for science, but maritime navigation, trade, and warfare). So, for historical and practical reasons, we know a fair amount about the deep circulation of the North Atlantic compared to some other areas of the world ocean. A more scientific justification for this semi-fictitious AMOC construct is that it embodies just about everything we think is important in the thermohaline circulation. Three of the main centers of sinking water *globally* are aligned with the Atlantic basin (another is Antarctica's Ross Sea, aligned with the Pacific). Although the North Pacific today lacks a major region of deep water formation because its surface waters are too fresh, very recent research has uncovered the fascinating possibility of a thriving "PMOC" during the Pliocene – a warm climate epoch about 3 million years ago (Burls *et al.*, 2017; Ford *et al.*, 2019). The AMOC is like a microcosm of the global thermohaline circulation, and while we might have a decent handle on its structure, we are just beginning to scratch the surface of how its strength varies over time!

Another slight incongruity that we must address is that the mechanisms for the thermohaline circulation discussed so far – that is, in terms of buoyancy forcing – only explain why water sinks. If it is a *circulation*, how does deep water formed at high latitudes return to the surface and rejoin the poleward flow for another go-round? The conceptual model of the thermohaline circulation with a premium placed on high-latitude convection was built up in the mid-twentieth century by many of the same characters appearing in Chapter 6, including Stommel (Stommel, 1958; Stommel and Arons, 1959a, 1959b), but work many decades earlier by Johan Sandström proved fundamental in closing the loop. **Sandström's theorem** stated that a steady-state overturning circulation can be maintained only if there is a heating source *at depth* (Sandström, 1908). This is actually quite intuitive; warm surface water cannot rise since it is already at the top, but warmer (and therefore less dense or positively buoyant) water below the surface *can* rise. This is again analogous to the atmosphere's Hadley circulation, where the heating and produc-tion of buoyant fluid is occurring at "depth" (the bottom of the atmosphere). Water below the surface is warmed by mixing – specifically, warm temperatures near the surface being mixed downward – and a major source of energy for that mixing is the wind (another is tides). Here's where the distinction between the wind-driven and buoyancy-driven (i.e., thermohaline) ocean circulation gets even more ten-uous. There are broad regions of persistent upwelling, such as off the coast of Antarctica (just northward of where AABW is formed) and along the equator in all three basins, and such wind-driven upwelling can tap into deep water masses, transporting some of their mass upward. It is therefore impossible to cleanly dis-entangle these circulations, and hard to avoid concluding that they even influence one another. As Stefan Rahmstorf summarized the matter:

> There are thus two distinct forcing mechanisms, but not two separate circulations. Change the wind stress, and the thermohaline circulation will change; alter thermohaline forcing, and the wind-driven currents will also change. It is because of thermohaline forcing that wind-driven currents are relegated to the upper ocean – in unstratified water they would extend to the bottom. (Rahmstorf, 2003)

Giving a meaningful, global summary to the thermohaline circulation is still challenging in the early twenty-first century, in part because the physical ocean-ographic community is still carefully measuring and exploring how it all fits together, especially the connectivity between the different ocean basins. Another reason for this challenge is that, unlike most features of the wind-driven cir-culation, such as the great subtropical gyres, direct measurements of its veloc-ity are fleeting. This is a *slow* circulation! If one water parcel could be followed throughout its global trek along the thermohaline circulation, it might complete that journey after a thousand years (if ever). So, actual velocities associated with the thermohaline circulation are, in real observations, mostly buried amid the swift, wind-driven, and geostrophic surface currents and are difficult to come by at depth. Many have attempted to sketch a map of a global, conveyer-belt-like thermohaline circulation; hundreds can be found on the internet – all of them are speculative, many even defy logic. Another such map won't be published here.

In the 1990s, Arnold Gordon of the Lamont–Doherty Earth Observatory in New York and Bill Schmitz (1937–2016) of the Woods Hole Oceanographic Institution in Massachusetts turned those diagrams on their heads – literally. Recognizing that the Southern Ocean is where all three major ocean basins (Atlantic, Pacific, and Indian) are connected via the Antarctic Circumpolar Current (ACC), and that the three basins do have some key aspects in common, something like Figure 8.4 emerged after carefully studying the available global observations of deep circulation and water masses (Gordon, 1991; Schmitz, 1996). The basic structure of the layered water masses in the Atlantic sector can be verified with **hydrographic profiles** (i.e., measurements of temperature, salinity, and other quantities at various depths taken from a stationary ship). Two such profiles were taken from the *R/V Melville* in 1989 under the leadership of Lynne Talley of the Scripps Institution of Oceanography, who was serving as the chief scientist of a routine oceanographic cruise in the South Atlantic. The locations of the two hydrographic profiles are

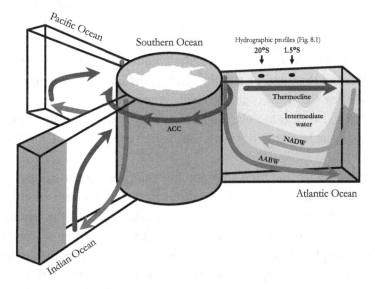

Figure 8.4 Schematic illustration of the global thermohaline circulation with an Antarctic-centered view of interbasin flows, inspired by the diagrams of Arnold Gordon (1991) and William J. Schmitz, Jr. (1996). Water mass details shown in the Atlantic sector include the thermocline, intermediate water, North Atlantic Deep Water (NADW), and Antarctic Bottom Water (AABW). The Antarctic Circumpolar Current (ACC) flows clockwise (from west to east) around Antarctica, connecting the three major ocean basins. The locations of the two hydrographic profiles shown in Figure 8.1 are labeled on top of the Atlantic sector of the diagram. (A black and white version of this figure will appear in some formats. For the color version, please refer to the plate section.)

shown on the map in Figure 8.3 and plotted on the T-S diagram in Figure 8.1. Near the surface, the two profiles, which are spaced by 18.5° latitude, start out quite different in terms of density (due to a large difference in salinity). Beneath the thermocline, however, the two hydrographic profiles converge on similar characteristics of Antarctic Intermediate Water (AAIW), NADW, and AABW. The details of the other two major ocean basins represented in the schematic illustration (Figure 8.4) are left intentionally simple; they are surely not so simple in the real world, but they do lack a significant source of sinking water in the high northern latitudes today. The Indian Ocean is, of course, abbreviated due to the southern edge of India and the rest of Asia around 10–20°N.

8.4   Climate Variability Linked to Buoyancy Forcing

Just like the Hadley circulation of the atmosphere, the importance of the buoyancy-driven, thermohaline circulation in shaping Earth's surface climate as we know it is clear. The overturning circulation accounts for an important fraction of the poleward heat transport necessary to balance the planetary energy budget (Chapter 5). In Chapter 7, we learned that the eastern equatorial Pacific Ocean is a vast, natural *source* of carbon dioxide to the atmosphere due to the upwelling of CO_2-rich water. In the high latitudes, including where the sinking flow (convection) of the thermohaline circulation occurs, the ocean is a CO_2 *sink* of equivalent magnitude. Being more soluble in cold water, CO_2 is drawn down from the atmosphere into the surface ocean, and the ongoing sinking relieves the build-up of pCO_2 at the surface so further air–sea CO_2 flux is maintained. When those molecules of greenhouse gases dissolve and sink at high latitudes, they may be temporarily sequestered into the deep ocean, at least for the long journey characteristic of the thermohaline circulation.

What happens when the buoyancy forcing changes? Historical and pre-historical records have shed light on past variations in surface climate, oftentimes attributing them to changes in the overall strength of the thermohaline circulation. This question is also raised in the context of anthropogenic (human-caused) climate change. As atmospheric carbon dioxide levels continue to rise, how will associated phenomena such as melting ice and tropical expansion affect the thermohaline circulation, and how might that feed back on greenhouse gas forcing itself and impact the geographic details of climate change over the coming decades to centuries? In the following sections, we will explore the interplay between climate variability and the thermohaline circulation. For now, we will look at these variations and relationships for what they are, or have revealed themselves to be thus far; we will save the specific problem of identifying the role of human activities for the next chapter.

8.4.1   The Paleoceanographic Perspective

The longest continuous thermometer record on Earth, from central England, began in the year 1659 CE. There are few direct temperature records extending even close to that far back in time, and coverage over the ocean is even more limited. We now have a reasonable handle on large-scale patterns of SST variability since the late 1800s, but they rely heavily on statistical methods and models to fill the gaps until the advent of Earth-observing satellites during the second half of the twentieth century. Fortunately, for those interested in a broader view on climate variability than our own life span, Earth's natural processes have accumulated a few time capsules recording climate variations over the ancient past. Some of them yield estimates of regional climate variability over the past few hundred years in places we otherwise had no information; others capture global climate cycles spanning millions of years. The scientific discipline concerned with generating and interpreting such **proxy** records of past climate is known as **paleoclimatology**. An overlapping branch of climate science that leans toward extracting such information from – and about – the ocean is called **paleoceanography**.

A paleoclimate proxy is something we are able to measure that is systematically *sensitive* to an actual climate variable, not a direct measure of that climate variable itself. Paleoclimate proxy *records* are a bit different than most of the data sets that have been used to illustrate the ocean–climate interactions so far in this book, and not just because they concern the pre-instrumental era. First, they are typically generated from a single *point* on Earth, rather than global *maps* that show how patterns change over time. The upshot, naturally, is that they provide a much longer-term perspective than maps of instrumental or satellite data. One may even be clever (or lucky) enough to extract a proxy record from a location that is representative of something bigger via large-scale patterns of climatic phenomena. Imagine a proxy record of SST from the eastern equatorial Pacific Ocean, for example; this might yield information on pre-historical El Niño events and not just SST variability at that single point. In general, among paleoclimate proxy records, there is a tradeoff between the *length* of its coverage and its *resolution* – the increment of time between each data point. A proxy record might be very long (millions of years), but give a relatively blurry view of the details within, while another might be shorter (hundreds of years) but with crisp resolution of climatic fluctuations from one season to the next.

One paleoclimate proxy in particular that has given us a wealth of insight about how Earth's climate has varied over long time scales is the **oxygen isotope fraction** ($\delta^{18}O$). Oxygen has three stable isotopes: ^{16}O, ^{17}O, and ^{18}O – each one differs from the next only in number of neutrons (as isotopes do). About 99.8 percent of all oxygen on Earth is ^{16}O (the superscript ^{16} is for 8 protons plus 8 neutrons inside the atomic nucleus). Another 0.2 percent of our oxygen is ^{18}O, with two extra neutrons (^{17}O only accounts for <0.1 percent and is not relevant here). In practice, the oxygen isotope fraction is defined relative to a known standard as follows, but remains directly proportional to the ratio of ^{18}O to ^{16}O of the sample of substance being measured:

$$\delta^{18}O = \left(\frac{\left(\frac{^{18}O}{^{16}O} \right)_{sample}}{\left(\frac{^{18}O}{^{16}O} \right)_{standard}} - 1 \right) \times 1000 \propto \frac{^{18}O}{^{16}O} \tag{8.9}$$

The $\delta^{18}O$ ratio found in certain substances is sensitive to climatic conditions, and it works as follows (Figure 8.5). When liquid water evaporates from the ocean, H_2O molecules composed of the "lighter" isotope of oxygen (^{16}O) are more likely to be transferred to the atmosphere as water vapor. This means that whenever evaporation occurs, the ratio of ^{18}O to ^{16}O, or $\delta^{18}O$, in the seawater is increased, while the $\delta^{18}O$ of the water content in the atmosphere is decreased. The opposite is true for precipitation; in that case, the "heavier" isotope (^{18}O) is preferred. By affecting the relative abundance of isotopes, evaporation and precipitation constitute mechanisms for oxygen **isotopic fractionation**. Both evaporation *and* precipitation serve to *increase* the $\delta^{18}O$ of seawater and *decrease* the $\delta^{18}O$ of water held in the atmosphere.

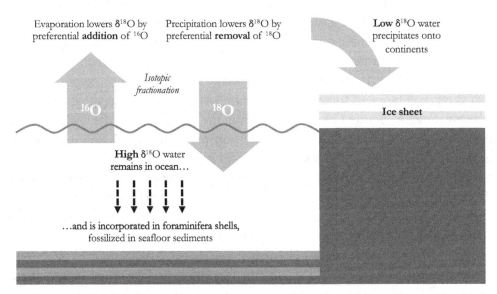

Figure 8.5 Schematic illustration of the mechanisms leading to oxygen isotopic fractionation. Evaporation and precipitation lower the $\delta^{18}O$ content of water in the atmosphere (which subsequently falls on land) and raise the $\delta^{18}O$ content of seawater (which is subsequently incorporated in foraminifera shells buried in layers of seafloor sediment). Cold climate conditions with large global ice volume are thus recorded as high $\delta^{18}O$ in benthic (seafloor) records and low $\delta^{18}O$ in ice core records.

Here's where $\delta^{18}O$ becomes a geochemical *proxy* for climate. Let's think about the "final" destination for all of this oxygen, or at least where paleoclimatologists go looking for it today. In the ocean, single-celled organisms called **foraminifera** build their shells using elements available in the seawater, including oxygen. When they die, their shells rain down onto the seafloor and become fossilized in the chronologically ordered layers of sediment. On land, water vapor condenses and falls as rain or snow (having been depleted of ^{18}O by several evaporation-precipitation cycles prior to reaching land). If the climate is warm enough that the fallen precipitation eventually flows back into the sea, then there is no signal to be recorded by seafloor sediments. At high latitudes or in otherwise cold climates, however, the fallen low-$\delta^{18}O$ water will *remain* on land as the next layer of an ice sheet such as those on Greenland and Antarctica. This means that when the global climate is cold, the ocean's $\delta^{18}O$ will be high, because much of the low-$\delta^{18}O$ water is locked in massive ice sheets on land, and vice versa. So, the $\delta^{18}O$ content of fossilized foraminifera in a layer of seafloor sediment is an excellent indicator of the global climatic conditions during the time they were alive and building their shells, particularly the global volume of ice. Long, cylindrical "cores" of sediment material are carefully extracted from the seafloor using drills lowered by ships; the further down, the longer back in time. This is such an enticing prospect that paleoceanographers have gone on a veritable drilling spree around the world ocean, generating one **benthic** (of the seafloor) $\delta^{18}O$ record after another.

In 2004, paleoceanographers Lorraine Lisiecki and Maureen Raymo produced a composite *global* benthic $\delta^{18}O$ record by merging 57 individual $\delta^{18}O$ records from around the world ocean (Lisiecki and Raymo, 2005). Such records are of the "long but blurry" type; seafloor sedimentation rates vary, but a temporal resolution (increment of time between data points) of about 1000 years is typical. The result is a stunning perspective on Earth's climate spanning more than five million years (Figure 8.6). The interglacial period (relatively warm and low global ice volume) that we live in today is just one of dozens that have occurred over the past few million years since the onset of the Pleistocene epoch (also known as the "ice age"). $\delta^{18}O$ records such as these are also part of how we came to be convinced that the glacial cycles that have repeated themselves over the Ice Age are linked to **orbital forcing**. Orbital forcing, also known as the Milankovitch cycles, includes variations in the eccentricity or "out-of-roundness" of Earth's orbit around the Sun (~100,000-year cycle), the obliquity or "tilt" of Earth's axis (~41,000-year cycle), and the precession or "wobble" of Earth's axis (~21,000-year cycle). Together, these three orbital parameters modulate the distribution of solar radiation cast upon Earth. A groundbreaking paper published in 1976 by Hays, Imbrie, and Shackleton (great-nephew of storied Antarctic explorer Ernest Shackleton) used similar – albeit fewer and shorter – $\delta^{18}O$ records to detect the strong effect of all three orbital parameters on global climate through the ice age

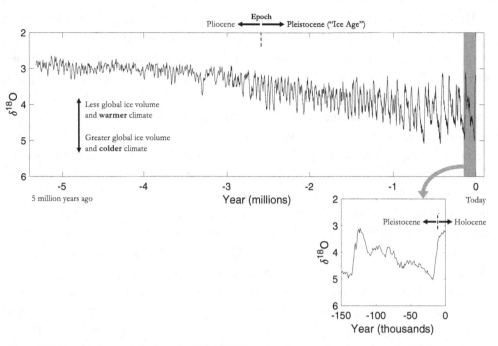

Figure 8.6 Record of oxygen isotope fraction ($\delta^{18}O$) from seafloor sediment cores, reflecting general variations in global climate including global ice volume over the past 5.3 million years. Note the reversed y axis; lesser values of $\delta^{18}O$ (higher on the graph) represent warmer global climate and lesser ice volume, and greater values of $\delta^{18}O$ (lower on the graph) represent colder global climate and greater ice volume. The record is a composite of 57 globally distributed records produced by Lisiecki and Raymo (2005). The inset zooms into the last 150,000 years, revealing the most recent glacial cycle and the boundary between the Pleistocene and Holocene (present) epochs at ~11,700 years ago.

(Hays *et al.*, 1976). It is simultaneously amazing and exciting that, despite the clear dominance of the glacial cycles evident over the past few million years of Earth's climatic history, we still have a long way to go to fully explain just how the variations in solar radiation associated with the three orbital cycles actually drives them. One thing is clear: The solar forcing changes alone are not enough to explain wholesale changes like 10 °C temperature swings between glacial and interglacial periods, so feedbacks play an important role in the periodic ice sheet growth and retreat. Feedbacks involving the carbon cycle and surface albedo are inevitable, but the role of the thermohaline ocean circulation remains mysterious.

On the other hand, there are abundant hypotheses and lines of evidence linking changes in buoyancy forcing and the thermohaline circulation to more abrupt climate changes recurring *within* the glacial cycles. One of our best views of climate variability over the last glacial cycle – since the previous interglacial – comes from records including $\delta^{18}O$ from ice cores drilled out of the Greenland ice sheet (Figure 8.7). Notice the level of detail present in the Greenland ice core $\delta^{18}O$ record compared to that of the global benthic $\delta^{18}O$ record (inset of Figure 8.6); due to relatively high ice accumulation rates (i.e., annual precipitation) on Greenland, each data point on Figure 8.7 represents a 50-year average. Then, notice the remarkable

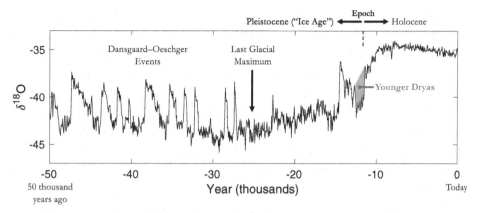

Figure 8.7 Record of oxygen isotope fraction ($\delta^{18}O$) from the North Greenland Ice Core Project (NGRIP), reflecting general variations in Northern Hemisphere climate, including the response to global ice volume over the past 123,000 years (only the last 50,000 years are shown here). Greater values of $\delta^{18}O$ (higher on the graph) represent warmer Northern Hemisphere climate and lesser global ice volume, and lesser values of $\delta^{18}O$ (lower on the graph) represent colder Northern Hemisphere climate and greater global ice volume. Note the greater detail in the ice core record than in the equivalent time period from the seafloor sediment cores (rightmost one-third of the inset of Figure 8.6). Dansgaard–Oeschger events, the Last Glacial Maximum (~26,500 years ago) and the Younger Dryas are labeled.

abundance of abrupt climate shifts that occurred during the last glacial period and during the jagged ascent into the present interglacial (i.e., the Holocene epoch). The sequence of 10 or so rapid coolings followed by even-more-rapid warmings evident in the Greenland $\delta^{18}O$ record between 50,000 and 27,000 years ago (and 25 in total during the entirety of the last glacial cycle) are known as **Dansgaard–Oeschger events**, and have been linked to the thermohaline circulation as follows. Massive ice sheets such as the Laurentide rimming the Arctic during the last glacial period occasionally grew to unstable dimensions, at which point purges of ice broke free (~10 percent of the Laurentide ice sheet) and spilled large volumes of freshwater into the ocean (recall that freshly melted ice is relatively devoid of salt). Some of these ice-breakoff events are called **Heinrich events**; evidence for them is manifest as fields of coarse debris buried in seafloor sediments far from continents, where the only reasonable explanation is that they were carried, or "rafted," by icebergs until the ice melted and deposited them on the seafloor. There is also evidence for concurrent slowdowns of the AMOC, which is reasonable considering the input of freshwater would increase the buoyancy of surface water and inhibit convection. Lending further support for a role of the AMOC in driving the Dansgaard–Oeschger cooling events evident in Greenland ice cores are nearly simultaneous warmings found in proxy records from the other side of the world – Antarctica. As a dump of freshwater into the Northern Hemisphere high-latitude oceans (especially the North Atlantic) weakens the AMOC, less heat is transported from the Southern to the Northern Hemisphere and results in something of a bi-polar seesaw (i.e., the two poles are out of phase – as one warms, the other cools).

Comparable mechanisms have also been offered to explain arguably the most abrupt climate shift of the last glacial period – one that occurred just as the climate system was digging out from the Last Glacial Maximum (LGM, roughly 27,000 years ago). Beginning around 12,800 years ago, the **Younger Dryas** event was marked by a ~5 °C cooling over a couple of decades and persisted for around 1000 years. The details remain debated, but one of the leading hypotheses involves another influx of freshwater to the high-latitude North Atlantic. Unlike the "binge-purge" cycles of the ice sheets driving the Dansgaard–Oeschger events, the pulse of meltwater this time was drainage from melting of the ice sheet as Earth's climate was ascending into today's interglacial period. Lake Agassiz was one of the enormous lakes left behind by glacial meltwater pooling in landscape depressions, just northwest of the remaining Great Lakes of the northern USA and southern Canada. As the landscape continued to evolve during the last deglaciation, Lake Agassiz drained into the North Atlantic via the Saint Lawrence River (and possibly into the Arctic/North Pacific via the Mackenzie River system), reducing convection associated with the thermohaline circulation and leading to a pronounced Northern Hemisphere cooling. After roughly 1000 years, the buoyancy-driven ocean circulation returned to vigor and the Northern Hemisphere climate quickly caught back up with the ongoing deglaciation by warming ~10 °C in as little as a few years!

What sets the abrupt climate shifts that occurred within the last glacial cycle (e.g., Dansgaard–Oeschger events and the Younger Dryas) apart from the glacial cycles themselves? For one thing, they are driven primarily by processes *internal* to Earth's climate system, whereas the glacial cycles with a period of ~100,000 years are fundamentally propelled by *external* forcing – variations in the distribution of solar radiation reaching Earth. Another key distinction, and one that brings a clear connection to the thermohaline circulation within reach, is that these abrupt shifts did not involve identical changes across the hemispheres, whereas the glacial cycles involved generally synchronized changes – cooling and warming together – from pole to pole. Given the asymmetry of the thermohaline circulation, and especially the AMOC, where the heat transport is *northward* regardless of hemisphere (Figure 8.4), it is quite natural to invoke regime shifts in the buoyancy-driven ocean circulation to explain the abrupt shifts focused in the Northern Hemisphere within the last glacial period, and far less straightforward for the glacial cycles themselves. Different AMOC *regimes*, or configurations of the circulation in terms of shape and strength, indeed may have existed at different times during the last glacial period, and one way in which they differ is the *latitude* at which convection (or NADW formation) occurs. While it may be necessary to suppose a wholesale shutdown of the AMOC during the most extreme Heinrich events, a mere southward shift in the latitude of convection from its current realm (e.g., the Nordic Seas) to the open North Atlantic, south of Iceland, just may have been enough to explain the high-latitude Northern Hemisphere cooling associated with most of the Dansgaard–Oeschger events.

Compared to $\delta^{18}O$ records from ice cores and, especially, seafloor sediment cores, paleoclimate proxy records extracted from **corals** provide a relatively high-resolution but necessarily shorter glimpse into climate variability. The paleoceanography community has developed a number of **paleothermometers** based on physical characteristics of – and material incorporated into – corals (again, extracted as cylindrical cores using drills). Most corals grow in diameter every year, and the visible striping of annual growth bands enables us to resolve climate cycles spanning years (rather than thousands of years). Analogous to tree rings, the variable width of annual growth bands of a coral (as measured by x-ray equipment) yields information about climate conditions. In general, the annual extension of a coral is inversely proportional to temperature (less growth during years of warm SST). Geochemical proxies derived from corals are typically defined as ratios of elements like magnesium (Mg) and strontium (Sr) to calcium (Ca), part of calcium carbonate ($CaCO_3$) – the essential ingredient of coral skeletons. Sr/Ca and Mg/Ca ratios are also sensitive to temperature through biologically and chemically mediated processes. Coral-based proxies therefore extend backward in time our records of natural climate variability in terms of SST, as discussed in the previous chapter.

Coral-based proxy records have been especially useful for understanding the Atlantic Multidecadal Oscillation (AMO), considering its time scale relative to the total length of instrumental records. Some such proxy records have revealed that the AMO is indeed oscillatory and several full cycles can be documented well before (and readily distinguishable from) the signal of strong anthropogenic warming (Vásquez-Bedoya et al., 2012). Long computer model simulations of the *coupled* climate system (i.e., models that include fully interactive atmospheres and oceans) have led to hypotheses linking the 60–80-year period of the AMO to a natural oscillation in the strength of the AMOC, which might work as follows (Vellinga and Wu, 2004; Knight et al., 2005). A strong AMOC leads to greater northward heat transport (manifesting as warmer SST in the North Atlantic – the very definition of the positive phase of the AMO) and consequently a northward-shifted and strengthened ITCZ. The additional freshwater flux (rain) to the surface ocean beneath the ITCZ lowers salinity, which is slowly propagated northward over the course of perhaps a decade, where its positive buoyancy intrudes on the convection residing at high latitudes. As a result, the AMOC weakens, the North Atlantic cools (equivalent to a negative AMO phase), and the ITCZ weakens and shifts southward. Thus begins the chain of events leading to the opposite phase of the AMOC cycle – a truly *coupled* mechanism (requiring interaction between the ocean and atmosphere). This is only one hypothesis for the AMO, but it is a great example of paleoclimate proxy records and models working together to test hypotheses about climate variability that are otherwise difficult to confirm with short records of surface climate and ocean circulation.

8.4.2   Modern Observations

In 2004, we entered an era of unprecedented observation of the thermohaline circulation. A legacy of the joint UK–USA project known as RAPID is a line of moorings across the Atlantic Ocean (along 26.5°N latitude) that measure density (from temperature and salinity) and then leverage geostrophic balance to compute the meridional volume and heat transports associated with the AMOC. Measurements by the RAPID Array thus far (Figure 8.8) reveal an average AMOC strength of ~17 Sv, with a seasonal cycle of about 10 Sv and a significant shift in 2008, when the overturning circulation strength dropped by 2.7 Sv (or 14 percent). The need for longer-term records (e.g., paleoclimate proxies) and computer models in order to connect such ocean circulation changes with climate is obvious, but these observations are a profound achievement by the international scientific community. They have certainly presented a host of new questions. Was the reduction in AMOC strength that took place in 2008 just a routine undulation associated with the AMO? Was it part of an ongoing, jagged trend associated with global warming? One and a half decades is not long enough to answer such intriguing questions with any certainty; we can only say that the change in AMOC strength happened. To put the RAPID observational record and the 2008 AMOC shift into a modestly longer-term context, it would be difficult to argue that a declining AMOC is *inconsistent* with other, ongoing changes in the climate system. The planet is warming, Greenland's ice (and Arctic sea ice) is melting into the North Atlantic, and a robust *cooling* trend is even emerging in the very spot on the surface ocean toward which the AMOC transports heat and dense water sinks (Figure 8.9). It almost seems like we've been here before – Dansgaard–Oeschger events and the Younger Dryas come to mind as possible glacial analogs of the events playing out in modern observations. It may be that only time will tell, but while we wait let's next lay out the dynamics of an evolving ocean in the twenty-first century.

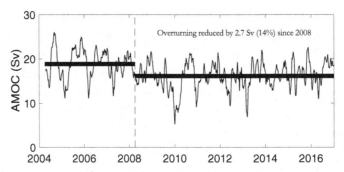

Figure 8.8 Time series of the observed strength of the Atlantic Meridional Overturning Circulation (AMOC), the part of the thermohaline circulation in the Atlantic Ocean, from transport measurements taken by the RAPID Array from April 2004 through February 2017. The thin line shows the 30-day smoothed data (original data set is 12-hourly resolution) and the thick lines show the averages before and after March 31, 2008.

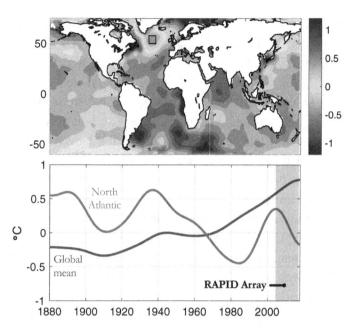

Figure 8.9 Map of linear trend in SST over the period 1880–2018 (°C per century). Time series of SST anomaly in the North Atlantic Ocean (averaged over the small box indicated on the trend map) (blue line) and global mean surface temperature anomaly (red line). Both time series are smoothed with a 30-year low-pass filter. SST data from the NOAA Extended Reconstructed SST version 5 (ERSSTv5) data set and global mean surface temperature from the NASA GISS Surface Temperature Analysis (GISTEMP) data set. All temperature anomalies relative to the 1951–1980 base period. (A black and white version of this figure will appear in some formats. For the color version, please refer to the plate section.)

Further Reading

See Rahmstorf (2002) for a review of the role of ocean circulation including the thermohaline circulation in climate variability over roughly the last glacial cycle (120,000 years).

Questions

1. Briefly explain why the mixing of two water masses of equal density can result in a water mass of greater density. What is this process called?
2. While on a research cruise, you observe the following three water masses at the surface. Based on the T–S diagram (Figure 8.1), which water mass is most likely to remain at the surface (as opposed to sink)?

    A. $T = 10\,°C$, $S = 35$ g/kg
    B. $T = 15\,°C$, $S = 36$ g/kg
    C. $T = 20\,°C$, $S = 37$ g/kg

3. The near-surface properties of the two hydrographic profiles taken along WOCE Line A16C in the South Atlantic Ocean in 1989 are substantially different, such that the surface density at $20°S$ is greater than that at $1.5°S$ by more than 1 kg/m^3 (see Figures 8.1 and 8.3). (a) Which variable that density depends on (T or S) is primarily responsible for the difference in density between these two stations, in this case? (b) Can you account for the change in T or S (depending on your answer to part (a)) between these two stations using your knowledge of the atmospheric general circulation (Chapter 5)?

4. Which two budgets described previously in this book are brought together by the density flux construct?

5. Examine the maps of density flux in Figure 8.2; notice the net negative density flux F_ρ along the West African coastline (~10–30°N). Considering the relative signs and magnitudes of the αF_T and βF_S components, form a hypothesis that invokes specific climatic processes to explain the net negative density flux F_ρ in this region.

6. Where does "deep water formation" occur in the ocean? Give a very general description, expressed as a function of both latitude and depth.

7. Compare and contrast the dominant physical processes leading to deep water formation in the Northern and Southern Hemispheres (e.g., NADW vs. AABW).

8. Briefly, what is the typical scientific tradeoff between a paleoclimate proxy and a global, gridded data set?

9. Heinrich events have been used as evidence to support a role for the AMOC in what type of abrupt climate change event in the paleoclimate record?

10. There are many sources of uncertainty associated with measuring the AMOC directly and detecting changes in its total transport over time (e.g., the RAPID Array). Specify one of them, and suggest how that particular source of uncertainty could be reduced.

11. The relative isotopic abundance of which element yields proxy information on the global ice volume? Explain why in 140 or fewer characters.

12. Select one of the *Dive into the Data* boxes featured in this chapter. Read the sample journal article that uses the associated data set, and describe the general role of that data set in the study. Provide a little context, both technical and scientific. What did the authors actually "do" with the data set? What specific scientific insight was enabled by incorporating this data set into the study?

13. Utilizing *Dive into the Data* Box 8.1, examine the vertical profiles of oxygen at the two stations. (a) How do they compare to one another? Specifically, in what depth range do they differ by the greatest amount? (b) Where is this region of greatest oxygen difference between the two stations in terms of temperature? (c) Speculate on the cause of the feature you are describing in parts (a) and (b), and why it is only present at one of the stations.

14. Utilizing *Dive into the Data* Boxes 8.4 and 8.5, examine the time series of AMOC transport (the smoothed version). (a) In what month and year did the strongest negative departure occur? (b) Produce a map of SST anomaly during that month, and plot a line representing the location of the RAPID Array. Comment on the pattern of SST anomaly in the Atlantic basin, especially relative to the RAPID Array measurements. (c) Design an index that can quantify the temporal variability of whatever you found interesting in part (b), and plot it over time. How does the anomaly in the month you identified in part (a) compare to the full record of SST beginning in 1854?

DIVE INTO THE DATA BOX 8.1

Name World Ocean Circulation Experiment (WOCE) Hydrographic Profiles

Synopsis The WOCE Hydrographic Program (WHP) has archived hydrographic observations (pressure/depth, temperature, salinity, and oxygen) taken during dozens of one-time cruises (sections) spanning the four major ocean basins.

Science Analysis of single hydrographic profiles enables identification of water masses when plotted in temperature–salinity space (*T–S* diagrams). When multiple profiles are combined into cross-sections, tracking of water masses and inference of ocean circulation is possible. Collections of such observations like the WHP enables a great deal of basic research on the ocean's structure and circulation.

Figure 8.1

Version 3

Variable Hydrographic properties (e.g., pressure, temperature, salinity, O_2) from lowered CTD sensor

Platform Ship

Spatial Sections (lines of vertical profiles, spaced every few degrees latitude or longitude), usually to near the seafloor (thousands of meters), throughout the Atlantic, Indian, Pacific, and Southern Oceans

Temporal 1985–1997, irregular intervals

Source www.nodc.noaa.gov/woce/woce_v3/wocedata_1/whp/onetime.htm

Format Plain text (.txt), ASCII (.csv), and NetCDF (.nc)

Resource woce.m; woce.mat (sample data and code provided on publisher website here: www.cambridge.org/karnauskas)

Journal Reference for Data

N/A, see www.nodc.noaa.gov/woce/woce_v3/wocedata_1/whp/manuals.htm

DIVE INTO THE DATA BOX 8.1 (cont.)

Sample Journal Article Using Data

Kouketsu, S., Kaneko, I., Kawano, T., *et al*. Changes of North Pacific
Intermediate Water properties in the subtropical gyre. *Geophys. Res. Lett.*
34, L02605 (2007). DOI: 10.1029/2006GL028499.

DIVE INTO THE DATA BOX 8.2

Name	Global Benthic $\delta^{18}O$ Stack
Synopsis	Merging of 57 globally distributed benthic (from the seafloor) $\delta^{18}O$ records. $\delta^{18}O$ records extracted from seafloor sediment cores represent global ice volume. The merged record ("stack") spans over five million years. This record was produced by Lorraine Lisiecki and Maureen Raymo in the early 2000s.
Science	Long records of climate such as proxies for global ice volume enable documentation and analysis of changes in global climate prior to the instrumental (and human) era. Gradual changes over the Pliocene and Pleistocene epochs as well as glacial cycles can be analyzed in the benthic $\delta^{18}O$ stack. Such records can be combined with those of orbital geometry (Milankovitch cycles) to infer the causes of ice age cycles.
Figure	8.6
Version	N/A
Variable	Oxygen isotope fraction
Platform	Sediment core
Spatial	1D (time series), global
Temporal	5,320,000 BCE to present, spacing between 1000 and 5000 years
Source	http://lorraine-lisiecki.com/stack.html
Format	Plain text (.txt)
Resource	benthic.m; benthic.mat (sample data and code provided on publisher website here: www.cambridge.org/karnauskas)

Journal Reference for Data

Lisiecki, L. E. and Raymo, M. E. A Pliocene–Pleistocene stack of 57 globally
distributed benthic $\delta^{18}O$ records. *Paleoceanography* **20**, PA1003 (2005). DOI:
10.1029/2004PA001071.

Sample Journal Article Using Data

Past Interglacials Working Group of PAGES. Interglacials of the last 800,000
years. *Rev. Geophys.* **54**, 162–219 (2016). DOI: 10.1002/2015RG000482.

DIVE INTO THE DATA BOX 8.3

Name North Greenland Ice Core Project (NGRIP) $\delta^{18}O$ Record
Synopsis Record of $\delta^{18}O$ from a 3 km long ice core drilled out of North Greenland (75.1°N, 42.3°W) between 1996 and 2003. The NGRIP ice core record extends back in time throughout the last glaciation and into the end of the last interglacial period.
Science Ice core records are shorter but of much higher temporal resolution than benthic records (sediment cores, e.g., the global $\delta^{18}O$ stack of Lisiecki and Raymo). The NGRIP record enabled improved documentation of the Last Glacial Maximum, study of abrupt climate changes during the last glacial period (e.g., Dansgaard–Oeschger and Heinrich events), and the evolution of Earth's climate during the deglacial transition into the present interglacial period (e.g., the Younger Dryas event). Further insights into abrupt climate changes and transitions, particularly their mechanistic linkages to global climate, also enabled by other measurements from the ice core record such as concentration of dusts (related to dryness) and sea salt (related to wind).
Figure 8.7
Version N/A
Variable Oxygen isotope fraction
Platform Ice core
Spatial 1D (time series), Greenland
Temporal 120,950 BCE to 2000 CE, 50-year averages
Source ftp://ftp.ncdc.noaa.gov/pub/data/paleo/icecore/greenland/summit/ngrip/isotopes/ngrip-d18o-50yr.txt
Format Plain text (.txt)
Resource ngrip.m; ngrip.mat (sample data and code provided on publisher website here: www.cambridge.org/karnauskas)

Journal Reference for Data

North Greenland Ice Core Project members. High-resolution record of Northern Hemisphere climate extending into the last interglacial period. *Nature* **431**, 147–151 (2004). DOI: 10.1038/nature02805.

Sample Journal Article Using Data

EPICA Community Members.One-to-one coupling of glacial climate variability in Greenland and Antarctica. *Nature* **444**, 195–198 (2006). DOI: 10.1038/nature05301.

DIVE INTO THE DATA BOX 8.4

Name	RAPID Array
Synopsis	Observed estimates of the volume transports of the Atlantic Meridional Overturning Circulation (AMOC) from moorings deployed across the North Atlantic Ocean, along latitude 26.5°N. The total AMOC transport is decomposed into various components including the Gulf Stream, Ekman transport, and transports in various depth layers. Rapid Climate Change–Meridional Overturning Circulation and Heatflux Array (RAPID/MOCHA) is a collaborative program between oceanographic institutions in the UK and USA.
Science	The RAPID Array enables careful quantification of the strength of the AMOC and how it varies over a range of time scales from days to years, and potentially trends and shifts in strength. Combined with related measurements, not only the volume transport but the heat transport can be quantified, which gives a better estimate of the ocean's contribution to the overall poleward heat transport required to balance Earth's latitudinal energy balance (between the tropics and poles). As the RAPID record grows longer, we may have better insights into the impact of anthropogenic climate change on the thermohaline circulation – a potentially severe change in global climate that models exhibit significant disagreement over.
Figure	8.8
Version	Release 2017
Variable	Volume transport
Platform	Mooring
Spatial	1D (time series), North Atlantic Ocean (along 26.5°N)
Temporal	April 2004 to present, 12-hour resolution
Source	www.rapid.ac.uk/rapidmoc/rapid_data/datadl.php
Format	ASCII (.ascii), MATLAB (.mat), and NetCDF (.nc)
Resource	rapid.m; rapid.mat (sample data and code provided on publisher website here: www.cambridge.org/karnauskas)

Journal Reference for Data

McCarthy, G. D., Smeed, D. A., Johns, W. E., *et al*. Measuring the Atlantic Meridional Overturning Circulation at 26°N. *Prog. Oceanogr.* **130**, 91–111 (2015). DOI: 10.1016/j.pocean.2014.10.006.

DIVE INTO THE DATA BOX 8.4 (cont.)

Sample Journal Article Using Data

Smeed, D. A., Josey, S. A., Beaulieu, C., *et al.* The North Atlantic Ocean is in a state of reduced overturning. *Geophys. Res. Lett.* **45**, 1527–1533 (2018). DOI: 10.1002/2017GL076350.

DIVE INTO THE DATA BOX 8.5

Name	NOAA Extended Reconstructed SST (ERSST)
Synopsis	Instrumental reconstruction ("reconstruction" here means statistical methods of filling gaps in the data) of monthly SST maps over the global ocean since the mid-nineteenth century using a wide variety of sources of SST observations from volunteer observing ships to satellites in recent decades (e.g., NOAA OI v.2). The most recent version (v.5) also includes state-of-the-art autonomous platforms (e.g., Argo floats). Other institutions produce similar global SST data sets, including the Hadley Centre in the United Kingdom. Differences between SST reconstructions can arise due to differences in raw data included and/or differences in statistical reconstruction/interpolation methodology. Serves as the surface temperature over the ocean in the GISTEMP data set (see Box 7.1).
Science	With more than a century and a half of global SST observations compiled into a single, consistent monthly data set, ocean and climate variability from seasonal to decadal time scales along with the long-term trend can be studied. Modes of climate variability (e.g., ENSO, PDO, AMO) can be investigated, including historical events and how they may change over time. When combined with other, terrestrial data sets, the influence of ocean variability on, e.g., droughts can be investigated.
Figure	8.9
Version	5
Variable	Sea surface temperature
Platform	Satellite, ship, mooring, float
Spatial	Global (gridded), 2° resolution

DIVE INTO THE DATA BOX 8.5 (cont.)

Temporal 1854 to present, monthly averages
Source www.esrl.noaa.gov/psd/data/gridded/data.noaa.ersst.v5.html
Format NetCDF (.nc)
Resource ersst.m; ersst.mat (sample data and code provided on publisher website here: www.cambridge.org/karnauskas)

Journal Reference for Data

Huang, B., Thorne, P., Banzon, V., *et al*. Extended Reconstructed Sea Surface Temperature, Version 5 (ERSSTv5): upgrades, validations, and intercomparisons. *J. Climate* **30**, 8179–8205 (2017). DOI: 10.1175/JCLI-D-16-0836.1.

Sample Journal Article Using Data

Caesar, L., Rahmstorf, S., Robinson, A., Feuler, G., and Saba, V. Observed fingerprint of a weakening Atlantic Ocean overturning circulation. *Nature* **556**, 191–196 (2018). DOI: 10.1038/s41586-018-0006-5.

9 Climate Change and the Ocean

"Change is coming, whether you like it or not."

Greta Thunberg

9.1 A Primer on Anthropogenic Radiative Forcing

Returning to the theme of planetary selfies from Chapter 1, there are some numbers streaming in today that do not paint a healthy or sustainable portrait. In 2018, humans emitted into the atmosphere about 10 gigatons[1] (Gt) of carbon from fossil fuel use, equivalent to 37 Gt of CO_2 (Le Quéré *et al.*, 2018). This number describing global **emissions**, which was quantified by an international team of dozens of credentialed scientists and engineers representing both academia and government agencies, has relatively small uncertainty (±5 percent). It is a couple of percent higher than in 2017, which was a couple of percent higher than in 2016. Another annual number that we know quite well is the amount by which the global **concentration** of atmospheric CO_2 has increased. In 2018, the concentration of CO_2 in Earth's atmosphere grew by about 3 parts per million (ppm), which is three times the annual growth rate[2] in 1970.

It is worth taking a moment to appreciate the distinction between emission and concentration, where the former is the amount deposited each year and the latter is the account balance. The strength of the greenhouse effect (described below) at any given time is a function of the composition of the atmosphere at that time, so the physical climate system responds primarily to total *concentration* of greenhouse gases (e.g., 408.52 ppm of CO_2 in 2018). If annual CO_2 emissions were as high as they are but constant year after year, the CO_2 concentration might increase linearly. However, the emissions themselves are still increasing over time, so the concentration is growing exponentially. Therefore, not only is the global carbon cycle changing, the change is accelerating as we speak.

---

[1] One gigaton is equivalent to one billion metric tons. In the Global Carbon Budget 2018 (Le Quéré *et al.*, 2018), "All quantities are presented in units of gigatonnes of carbon (GtC, 10^{15} gC), which is the same as petagrams of carbon (PgC; Table 1). Units of gigatonnes of CO_2 (or billion tonnes of CO_2) used in policy are equal to 3.664 multiplied by the value in units of GtC."

[2] NOAA defines the annual CO_2 growth rate as the total change from the first day of the year to the last (not one annual average minus the last). See www.esrl.noaa.gov/gmd/ccgg/trends for more information.

The amount by which the atmospheric CO_2 concentration has increased since the beginning of the Industrial Revolution is not an unprecedented change (see Figure 1.6). Both the rise in CO_2 concentration over the past century and a half due to anthropogenic emissions and the increases in CO_2 that occur at the termination of a glacial cycle are on the order of 100 ppm. However, it is the pace of the recent rise in CO_2 concentration that is unprecedented in Earth's history as we know it (Figure 9.1). If we line up the beginning of the previous two deglaciations (roughly 136,900 and 15,400 years ago) with the year 1851, we can see just how rapidly CO_2 concentration is rising today, in comparison to those pivotal climatic shifts in Earth's history. The previous two deglaciations took between about 6,000 and 10,000 years to allow CO_2 concentration to rise by 80 ppm, whereas anthropogenic emissions drove CO_2 upward by the same amount in 150 years (1851–2000). At the present rate of increase (2.5 ppm per year), the next 80 ppm should be added in as little as 30 years. We are conducting a massive, uncontrolled physics experiment on planet Earth.

That human activities are emitting CO_2 and the atmospheric CO_2 concentration is rising is not a coincidence. Taken separately, they are highly uncontroversial facts both within and outside of the scientific community. There are a couple of **attribution** ("cause and effect") cases that must be made, however, and although science has made them very well, they are the usual sources of confusion and contention among those whose expertise and life's work may lie outside of the field of climate science. The first is whether CO_2 is rising at the measured rates

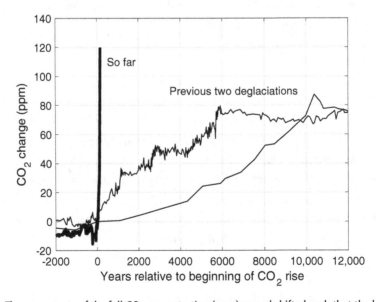

Figure 9.1 Three segments of the full CO_2 concentration (ppm) record shifted such that the beginning of the CO_2 rises associated with the previous two deglaciations and the modern era begin at coordinates (0,0) on the graph. This enables direct comparison of both the amount and rates of change associated with these three time periods. The last and next-to-last deglaciations were taken to start at years −15,410 and −136,900, respectively, and the modern era is aligned such that the year 1851 is at zero on the time axis. The data are identical to those displayed in Figure 1.6.

because of human activities. After all, Earth's climate, including the atmospheric CO_2 concentration, has changed in the distant past (a fact all-too-often proffered to climate scientists, as though they were not the ones who discovered it in the first place). For a solid case, we go beyond simply correlation, which – and pardon the cliché – doesn't prove causation.

Paleoclimatologists have an incredibly powerful (and now familiar) tool to detect the fingerprint of fossil fuel combustion: isotopic fractionation. When we burn fossil fuels, we are mostly burning long-dead and decayed plant matter that has been buried in layers beneath the surface of the Earth and fossilized (hence we call them "fossil" fuels, leaving young children bewildered why we'd do such a thing to dinosaur bones). When those plants were alive and photosynthesizing, they were fixing carbon in the form of atmospheric CO_2. However, not all carbon atoms are the same. There are three naturally occurring carbon isotopes, which differ only in how many neutrons are present (as all isotopes do). Carbon-12 (^{12}C) and carbon-13 (^{13}C) are stable, meaning they do not decay radioactively like carbon-14 (^{14}C). When plants photosynthesize, they may use either ^{12}C or ^{13}C, but their physiology is such that they are able to discriminate between carbon isotopes and they strongly prefer the lighter isotope, ^{12}C. This process of isotopic fractionation is akin to that described in the previous chapter for oxygen, leading to a way to deduce past variations in global ice volume. Therefore, tracking the ratio of one isotope to the other in the atmosphere ($^{13}C/^{12}C$ or simply $\delta^{13}C$) enables us to determine *why* CO_2 started rising so rapidly as it did in the nineteenth century. By examining air bubbles trapped in long ice cores that were extracted from the Antarctic ice sheet, studies since the late 1990s have indicated a precipitous drop in atmospheric $\delta^{13}C$ (higher ^{12}C relative to ^{13}C) timed very well with the increase in total atmospheric CO_2 concentration, confirming that the modern rise in CO_2 is attributable to fossil fuel combustion (Francey *et al.*, 1999; Rubino *et al.*, 2013). Have anthropogenic emissions been *enough* to account for the atmospheric concentration passing 400 ppm in 2015? Unfortunately, we've emitted way more than enough to account for that. Had all of the CO_2 emitted into the atmosphere *stayed* in the atmosphere rather than being absorbed by the ocean and land, we would have met the 400 ppm milestone about 20 years earlier (Denman and Brasseur, 2007). Half of the CO_2 emitted by fossil fuel combustion so far has already been drawn out of the atmosphere, and that burden has been shared about evenly between the ocean and terrestrial biosphere (Figure 9.2).

The second question of attribution concerns whether the rising CO_2 concentration (which is definitely due to human activities) is actually causing the observed climate changes including global warming. Even with the acceptance that the global average surface temperature on Earth has risen by about 1.3 °C (2.4 °F) over the last century, firmly establishing cause and effect here requires some basic understanding of the expected thermodynamic response of the climate system to increased greenhouse gas concentrations, ultimately building those laws into comprehensive Earth system models that allow for quantitative attribution of the observed

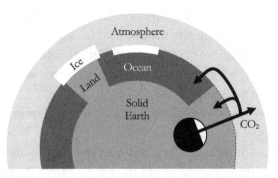

Figure 9.2 Illustration of carbon reservoirs and components of Earth's climate system, including the atmosphere, ocean, cryosphere, and solid Earth. Black arrows represent anthropogenic carbon dioxide emissions, indicated as a flux from the solid Earth into the atmosphere, and subsequent fluxes of anthropogenic CO_2 from the atmosphere to the terrestrial biosphere and ocean.

warming to the actual CO_2 emissions. Why would we expect, based on the laws of thermodynamics, the climate to warm when we add additional molecules of CO_2 to the atmosphere? Carbon dioxide is deemed a **greenhouse gas** because of the way it interacts with longwave radiation attempting to pass through it. If you imagine a molecule of CO_2 as three marbles (two oxygen atoms and one carbon in the middle) connected by two springs (covalent bonds), there are many ways for such a toy to bend, wiggle, and vibrate. Like other greenhouse gases, including methane, nitrous oxide, and water vapor, the geometry of these molecules is well suited to intercepting longwave radiation, converting it to kinetic energy (i.e., raising their own temperature), and re-emitting that longwave radiation at that temperature (where longwave emission is always proportional to T^4), just like in Chapter 2.

A greenhouse is therefore not a bad analogy for this process; shortwave radiation emitted by the Sun passes through these molecules in the atmosphere to heat the surface (like sunlight passing through glass walls), but most of the longwave radiation emitted by the surface is intercepted and re-emitted by greenhouse gases on its way out. If there were no greenhouse gases whatsoever in Earth's atmosphere, the planet's equilibrium temperature would be a frigid –18 °C. That is the temperature of a rock at a distance of 93 million miles from the Sun where 240 W/m^2 of shortwave radiation is exactly balanced by 240 W/m^2 of longwave radiation emitted back to space. It is perhaps obvious but nonetheless important to note that greenhouse gases are not just lining the floor; they are more or less mixed throughout the depth of the troposphere. So, when greenhouse gases absorb and re-emit longwave radiation, the fraction of that re-emitted energy that is escaping to space has been re-emitted on average by molecules with a *lower* temperature than that of the surface because they reside in a colder part of the atmosphere! It is both the radiative properties of greenhouse gases *and* their vertical distribution in the troposphere that render them very effective at reducing the amount of longwave radiation ultimately escaping to space. Since they do not change the amount of incoming (solar) radiation, the incoming radiation in the presence of greenhouse gases would exceed the outgoing radiation. Given such a net energy imbalance, the fundamental laws of thermodynamics for a closed system guarantee that a new equilibrium will be reached after the temperature increases until the outgoing radiation exactly balances the incoming radiation. Given the actual loading of greenhouse gases in Earth's atmosphere, this equilibrium is (or was) reached with the average surface temperature of Earth set at about 15 °C.

Every additional molecule of a greenhouse gas emitted to the atmosphere by human activity ensures that more of the longwave radiation emitted by Earth's surface is intercepted in the atmosphere and re-radiated at a lower temperature, thus maintaining a perpetual *disequilibrium* between the incoming and outgoing energy. This energy imbalance currently amounts to a **radiative forcing** of $1-2\,\mathrm{W/m^2}$, and meanwhile we witness the guaranteed consequence of an upward trend in global average surface temperature. This is not just theory. Observational studies have unambiguously detected the fingerprints of radiative forcing in measurements of the *decreasing* trend of longwave radiation escaping Earth (Harries *et al.*, 2001), and of the *increasing* trend of longwave radiation hitting the surface of Earth (Feldman *et al.*, 2015). The greenhouse effect works, and we are making it stronger.

As the greenhouse effect and the laws of thermodynamics in general are quite well known, they form an important element in the suite of global climate models that are discussed in the next section, and which we use both to attribute past changes in climate and to project future changes given various assumptions about how much CO_2 will be emitted throughout the remainder of this century. One area where those models really shine is enabling us to also anticipate regional manifestations of global warming and the cascade of other impacts that are mediated through warming, such as melting ice and sea level rise.

9.2   Numerical Ocean and Climate Models

9.2.1   What Is a Climate Model?

Section 9.3 will examine some of the manifestations and pathways by which climate change plays out in the world ocean, both past and future. For every such example, the observational platforms are different and ingenuitive in their own way. Some changes are measured by ships, others by Earth-orbiting satellites, and others still by remotely operated *in-situ* sensors. In most instances, however, there is a common tool that is used for projecting the future impacts and regional expressions of climate change: **global climate models**, or GCMs for short. Actually, the acronym GCM was traditionally meant to refer to *general circulation* models, and unless an "O" for ocean was placed at the front (as in OGCM) when spoken as such, it was just assumed to be strictly a model of the *atmospheric* general circulation. Times have changed, and now it is well-recognized that climate simulations without the ocean have limited applicability to the real world. In this chapter, we will use GCM to refer to global climate models – that is, comprehensive models of the global climate system that incorporate the atmosphere *and* ocean, as well as other realms of the Earth system like the cryosphere (including sea ice, ice sheets, snow, and glaciers) and even the biosphere to a varying extent. Modern GCMs are actually *combinations* of atmospheric general circulation models (AGCMs), ocean general circulation models (OGCMs), and other sub-models, like ones that predict how ice sheets and fields of sea ice change and how terrestrial vegetation responds to climate. Global climate models are also useful for understanding the past, as certain experiments can be designed to estimate how much *natural* variability we might expect to see in various aspects of the climate system, and to attribute observed long-term changes in the climate to particular forcings. Those experiments will be discussed here as well.

What are climate models? To the uninitiated, the answer to that question can take a little while to sink in, particularly because of how tempting it is to conflate them with climate observations. So, let's start by addressing the role of observations head-on. In the most general sense of the word, *observations* of the natural world have led us to theories that explain in a universal sense how it all works. Think of Newton's *Principia* of 1687. He observed the proverbial apple falling from the tree, and following much more careful observation and consideration, the laws of classical mechanics, gravitation, etc. were developed. Most of the equations and other theoretical material comprising a modern physics textbook were inspired at some point in time by a range of such observations that made someone ask "why?" The various models (of the atmosphere, ocean, etc.) that make up a complete GCM are actually nothing but the laws of physics concerning those realms of the climate system, all packed into a computer program that solves those equations at every location on Earth and on any date (past or future) that is of interest to the investigator. Notice that the "observations" that were important in the development of GCMs were actually important in the development of their precursors – the laws of physics. They are *not* observations of climate change. Climate change is the thing we want to diagnose and predict with GCMs, so observations *of* climate change are not fed into GCMs, and for some very good reasons that we will touch on below. To summarize, GCMs are large collections of equations whose outputs impact each other's inputs, and we know those equations because of insightful observation of the natural world that has occurred over centuries. In the case of an OGCM, or the ocean component of a GCM, the preceding chapters of this book (especially Chapters 2, 3, 4, 6, and 8) lay out the physics and dynamics that are at work in such a model – they are just then translated from mathematical notation into computer code.

9.2.2   The Structure of GCMs and Supercomputers

The above general description of GCMs may leave you wondering why we need supercomputers the size of large classrooms to run climate simulations with them.

The answer has less to do with the number of equations that comprise a GCM and more to do with how big the Earth is and how far away the year 2100 is. If you imagine all of the equations in the preceding chapters of this book, plus all the equations from books that similarly describe the behavior of the atmosphere, sea ice, ice sheets, and terrestrial biosphere – especially since we usually framed them in an Eulerian (local) perspective – you can now imagine how simulating the time evolution of the climate system actually means solving those equations *everywhere!* It should also be clear from some of the concepts in the preceding chapters that those equations at one location cannot just be considered in isolation from neighboring locations. Think about temperature advection, for example. The time rate of change of temperature $\partial T/\partial t$ at one location depends not only on the velocity $\mathbf{V}$ at that location, but also on the surrounding temperature gradient ∇T. On top of that, the new temperature T is a function of $\partial T/\partial t$ plus the previous time's temperature (so a single time step cannot be simulated in isolation from all of the previous time steps). It gets complicated and expensive quickly (since time on supercomputers is rarely free)! Fortunately, there is some compromise that takes place, which means we can approximate "everywhere" with a finite number of cubes.

Global climate models take the world and divide it into three-dimensional shapes called **grid cells** (Figure 9.3). If it is decided that each grid cell in the *ocean* component of the GCM is 1° latitude by 1° longitude by 100 m deep, we end up with around three million grid cells. These numbers define the **resolution** of the model. In reality, the resolution of a model can be different in different regions, depending on what the modeler thinks is going to be important to resolve in greater detail. For example, many OGCMs have "stretched" resolution such that grid cells are smaller (~0.25° latitude) but more numerous within a few degrees of the equator, and finer vertical resolution (1–10 m) in the upper ocean. The former is in recognition of how important ENSO is in the global climate system, and the latter of how detailed the dynamical processes in the upper ocean near the thermocline and mixed layer are, how crucial those processes are in determining sea surface temperature (SST), and how sensitive the atmosphere can be to SST anomalies (Section 5.5). The global atmospheric component of a GCM is also structured with a resolution that may vary both horizontally and vertically. Finally, an important component of a GCM known as the "coupler" serves to translate everything that is happening in the AGCM that should impact the ocean into *forcing* to the ocean (wind, solar radiation, etc.). At the same time – and this has to happen at a relatively frequent time interval during a coupled simulation – the coupler must also round up all of the variables from the OGCM that might impact the atmosphere (SST being the key one) and pass it along as bottom boundary forcing to the AGCM. So, a GCM is a set of multiple models that are each responsible for simulating their own realm, but exchanging information frequently along the way so that the overall climate evolution is determined by their *coupled* interaction – just like in the real world.

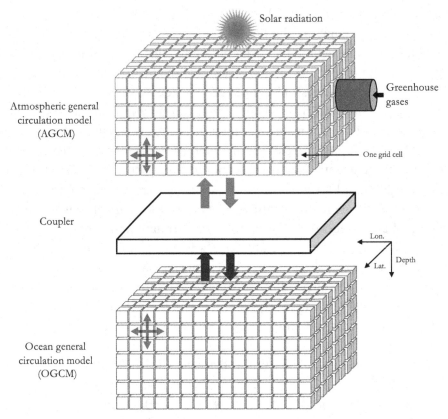

Figure 9.3 Anatomy of a GCM as a *coupled* model including component models representing the global atmosphere (AGCM), global ocean (OGCM), and other components such as ice and the terrestrial and/or marine biosphere (not pictured). The component models discretize their global realm with finite-sized grid cells whose size is a function of model resolution. Grid cells must communicate with each other (transparent arrows), while a coupler serves to exchange information (forcing) between the component models. Typical prescriptions during coupled GCM experiments include solar radiation and the concentration of greenhouse gases.

If we were talking about just a standalone AGCM or OGCM simulation, then the boundary forcing would normally be prescribed by observations (e.g., a gridded data set of surface wind stress observations forcing an OGCM from above, where those wind stress observations will not be altered by how the SST field is evolved by the OGCM). Since GCMs are *coupled* models, there is no way to guarantee that the basic climatology that emerges in such simulations is perfect. Each model has mean state **biases**, meaning a little too much precipitation here, not enough there, this part of the ocean is too warm, or not enough wind over there. Within reason such biases are acceptable, but it is still the concern of model developers to improve them with each new version of their GCMs.

The amount of computer time required for a given GCM to complete a simulation is known as "wall clock" time – this is how long you, the climate modeler, will actually have to wait for your simulation to finish and be able to make graphs of the results. Wall clock time depends on the following characteristics of the GCM and the experimental design: spatial resolution, time step (e.g., a new climate state is solved for every six hours), coupling frequency (e.g., the atmosphere and ocean exchange information every day), and the number of years to be simulated (e.g., 1850–2005 or 2006–2100). The wall clock time also depends on the computer! Drawing on the comparison between the size of a supercomputer and a large classroom, a supercomputer is not unlike a classroom full of students, each with their own basic calculator. The supercomputer takes the whole Earth full of grid cells as defined by the modeler, splits it up into a number of segments (think regions), and assigns each segment to an individual processor to work on – a computing strategy known as **parallel processing**. This way, the large number of calculations needed just for a single time step in the model simulation is shared across a (hopefully large) number of processors who work on them simultaneously. When the first time step is finished (say 0 UTC on January 1, 1900), all of the processors move on to the next time step (6:00 UTC), solve the equations for each grid cell in their assigned segment of the world, and so on. Eventually, 0 UTC on January 1, 2000 is reached and *voila*, a simulation of the last century has been completed. For state-of-the-art GCMs running on impressive supercomputers, such a simulation might take several days if not weeks or more, depending on the aforementioned characteristics of the GCM and experimental design.

9.2.3 Global Climate Models and Global Change Science

While running a one-off GCM experiment for an interesting theoretical study might be a solitary pursuit achievable by a single scientist, climate modeling for the purpose of global change science is a truly collaborative and international enterprise. There are some 50 GCMs that have been developed and are run routinely at various research, academic, and governmental institutions around the world. Major centers in the USA, for example, include the National Center for Atmospheric Research (NCAR), the NASA Goddard Institute for Space Sciences (GISS), and the NOAA Geophysical Fluid Dynamics Laboratory (GFDL). These are some incredibly sophisticated computer models; *within* each of these institutions, there are large and diverse teams of scientists (both climate scientists and computer scientists) whose full-time jobs are to work on a single component of the full GCM (like the OGCM, AGCM, ice models, etc.).

When it comes time to conduct the important sets of climate change experiments that are vital to the periodic Assessment Reports of the Intergovernmental Panel on Climate Change (IPCC), a team must be assembled to coordinate *across* institutions. This effort is called the Coupled Model Intercomparison Project (CMIP). Just as the latest Assessment Report of the IPCC was AR5 (IPCC, 2013), the most recent CMIP was CMIP5. The coordination achieved by CMIP is essential to ensure that, while each of the international modeling institutions may have their own GCM or GCMs, the experiments being conducted have exactly the same design. In other words, each GCM simulation uses the same prescribed historical greenhouse gas forcing, the same time periods being simulated, and the same assumptions about possible *future* CO_2 concentrations. The latter is of course not known and depends very much on what global society decides to do in the coming decades, but at least the same Representative Concentration Pathways (RCPs) are simulated by each model. This coordination and controlling for as many confounding variables as possible leads to a uniformity of experiments that enables CMIP and the IPCC authors to combine the results from all models into a single picture, ensuring that any differences between the results from different institutions must be the result of their *models themselves* being different, not, say, because of different estimates of past methane concentrations being used.

Several types of GCM experiments are conducted under the auspices of CMIP every time a new IPCC Assessment Report is on the horizon (every 5–7 years). **Control experiments** are GCM simulations in which the radiative forcing (greenhouse gases, volcanic aerosols, solar radiation) are held constant for hundreds to thousands of years. Control experiments yield insight into how much the climate system varies without any perturbation by humans or other external factors, which is important to know when determining whether a trend is undoubtedly due to radiative forcing or might just be a perfectly natural wiggle. **Attribution experiments** are two otherwise-identical GCM simulations carried out with and without "something," where that something is usually anthropogenic forcings like greenhouse gases. These are "what if" experiments, where we can estimate how the past century of climate would have unfolded *without* fossil fuel combustion simply by keeping the atmospheric CO_2 concentration set at its 1850 value. Here's why observations of climate change (beyond the *forcing* like CO_2 concentration) are not just fed into the models! Since one of these GCM simulations represents the model's best attempt to reproduce the past increases in global average temperature (and other associated consequences of anthropogenic radiative forcing) and the other excludes the greenhouse gases, we can *compare* both of them with the observational records and soundly establish attribution – we *cannot* explain the observed global warming unless we include the effect of greenhouse gases (Figure 9.4). It is not just a natural wiggle or due to some other natural forcing like volcanic aerosols or variations in the shortwave radiation emitted by the Sun. We humans are responsible for the rise in greenhouse gases, which explains the observed warming; therefore we are responsible for the observed warming – full circle.

Finally, **future experiments** are conducted by running GCMs from the present to the year 2100 (or beyond), and leaking CO_2 into the atmosphere according to an assumption of how much humans will continue emitting the gas over that period

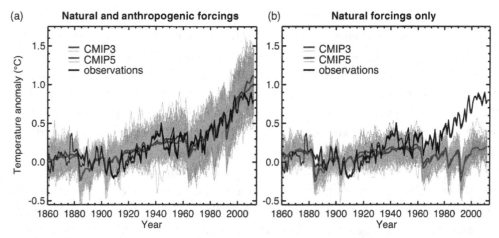

Figure 9.4 Historical simulations by ensembles of GCMs of the CMIP3 and CMIP5 generation of global mean surface temperature (°C) with natural (e.g., volcanic) and anthropogenic (e.g., greenhouse gases) forcings (a). The plot in (b) is the same except that only natural forcings were included in the model simulations, thus *attributing* the observed temperature change to anthropogenic forcings. Three different observational records are shown in black lines. Individual model simulations are shown in thin light blue and yellow lines, while the multi-model average is indicated by thick blue and red lines for CMIP3 and CMIP5, respectively. Adapted from IPCC AR5 WG1, Figure 10.1 (IPCC, 2013). (A black and white version of this figure will appear in some formats. For the color version, please refer to the plate section.)

of time. This is where the coordinating role of CMIP and IPCC is especially crucial. Rather than allowing every nation or modeling institution to use their own projection of CO_2 emissions, which is highly dependent on assumptions about population, technology, energy policy, international relations, national politics, etc., a set of four **Representative Concentration Pathways** (RCPs) were agreed upon by CMIP and IPCC stakeholders, and each modeling center used those same four RCPs so that, again, all of the model results could be combined onto a single graph (literally) to enable us to judge not only the consensus predictions but the range of uncertainty. The most severe RCP is RCP8.5 (where 8.5 refers to 8.5 W/m² of radiative forcing due to the climatic energy imbalance at year 2100), and was previewed in Chapter 1 (Figure 1.6). Representative Concentration Pathways of lesser end-of-century radiative forcing fall somewhere below RCP8.5 on that graph: RCP6 and RCP4.5 may still be reasonable scenarios, while RCP2.6 sits so close to the year 2018 concentration by 2100 that one can only imagine the magnitude and abruptness of the global societal changes required for it to become reality. In most peer-reviewed studies considering future climate changes, and in the Fifth IPCC Assessment Report itself, the most common comparison is between RCP8.5 and RCP4.5 as something of a comparison between unmitigated or "business-as-usual" and "policy-mitigated emissions" scenarios, although even the term business-as-usual makes strong (and actively debated) assumptions about what we are capable of.

While it has been the norm to express the results of future experiments by GCMs as the change of a variable (such as rainfall) at a particular time horizon in the future (such as 2050) given a specified forcing scenario (such as RCP4.5) and perhaps compare that with the same result but from the RCP8.5 simulations, a new and clever alternative has grown in popularity in recent years as a way to make the results more relevant to ongoing international climate policy discussions such as the 2 °C target associated with the Paris Agreement. Rather than a *time horizon*, results can be portrayed for a particular *global warming threshold*. For example, the resulting illustration might be a map of the rainfall change that we expect if and when global warming reaches 2 °C, which can be compared with the change expected for 1.5 °C and 3 °C to weigh the costs and benefits of meeting such global warming targets through mitigation strategies. While this might not satisfy the stakeholder who *is* interested in a particular time period, when averaging multiple models (as is usually the case in climate modeling for global change science), it does have the clear advantage of removing the confounding factor that every GCM might have a different **climate sensitivity** (how much the global average surface temperature changes for a given amount of greenhouse gas forcing). If a GCM developed and run by NASA GISS reached 2 °C global warming by 2030 under RCP8.5 forcing, but the one at NOAA GFDL didn't reach a global warming of 2 °C under the same forcing until 2070, does it make sense to pull out the climate at year 2050 to compare between both models? Again, it depends on whether the stakeholder using the GCM results is interested in adaptation measures that must be in place by 2050 or understanding the climate impacts that can be expected, given different levels of global warming.

Finally, simulations are also carried out to simulate paleoclimatic conditions. Paleoclimate experiments in particular provide an additional check on GCM reliability since they are simulating climatic conditions not experienced in the instrumental record (just like the future), yet in this case we *do* have some constraints on reality from the suite of paleoclimate proxies, as mentioned in the previous chapter.

9.2.4   Uncertainty in GCM Results

All of the coordination by CMIP and IPCC to ensure uniformity across climate change simulations carried out at modeling institutions around the world has paid off in many ways, but especially in terms of our ability to characterize the **uncertainties** in future climate change predictions. If the world only had one GCM, and we only ran one future experiment once, we would be in a situation even worse than having uncertainty: We would not even know what the uncertainty is. Although there are probably many segments of computer code that may be shared between GCMs or at least descended from a common ancestor, each GCM is a little different from the others. Quite logically, the attributes they share in common are the aspects of the climate system we understand the best (basic thermodynamic principles, basic force balances driving the wind and ocean currents, etc.), and where they differ the most are in those aspects that are either difficult to simulate or are heavily *parameterized* because we cannot explicitly resolve them. Examples of the latter include clouds and precipitation, vertical mixing in the ocean, diffusion, and boundary layer processes in the atmosphere. Simply put, they are driven by processes occurring at spatial (or temporal) scales smaller than the grid cells that constitute a GCM, so each model development team must employ their own innovations to represent them since they surely matter in the big picture. These challenges, either to our scientific understanding of the climate system or to our technical capabilities to run models of high enough resolution, has led to a useful diversity of GCMs. The same prescribed CO_2 forcing in 50 models results in 50 different climate simulations. This is what we call the **scientific uncertainty** associated with GCM simulations (Figure 9.5), and is what so many in the climate science enterprise (be it atmospheric science, physical oceanography, ice sheet dynamics, or a variety of other disciplines) work tirelessly to reduce, whether they are modelers or observationalists!

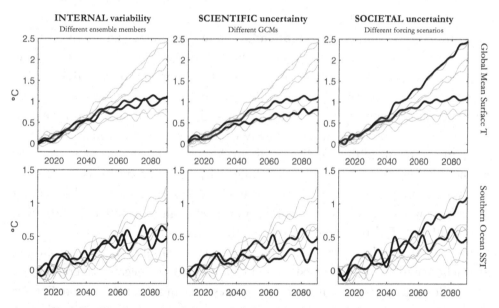

Figure 9.5 Schematic illustration of the major sources of uncertainty in future climate projections by GCMs. Top row for global mean surface temperature, bottom row for Southern Ocean SST. The same eight simulations are repeated across each row (2 GCMs, 2 RCPs, 2 ensemble members). The first column highlights uncertainty due to internal variability by contrasting two ensemble members of the same GCM and subject to the same future radiative forcing scenario (RCP4.5). The only difference between ensemble members is the initial conditions (weather on day one of the experiment). The second column highlights scientific uncertainty by comparing two different GCMs subject to the same forcing scenario (RCP4.5). The third column highlights societal uncertainty by contrasting projections by the same GCM but under two different forcing scenarios (RCP4.5 and RCP8.5).

There are two other important sources of uncertainty that are quantifiable, given the suite of model simulations associated with CMIP and reported on in IPCC Assessment Reports. If the scientific uncertainty is essentially the *spread* about the average of model results for the same RCP, the **societal uncertainty** is the opposite – the difference between the average results associated with two different RCPs. No matter how hard we try, we will always have a considerable amount of societal uncertainty. In other words, even if we had perfect GCMs, they would still predict different outcomes for different assumptions about future CO_2 emissions. The final major source of uncertainty is a bit tricky, and only became widely recognized as a major gap in our understanding of uncertainties in the past decade or so. As each *coupled* model simulation integrates from one year to the next, it is free to determine its own **internal variability**. In other words, an enormous El Niño event might happen in year 2037 in one model, but there is no reason such an event will occur in the same year in any other model simulation. This is because the only thing being prescribed in such simulations is the CO_2 forcing, and gradually trending CO_2 forcing doesn't cause the ENSO events. In general, internal variability is a greater impediment to identifying anthropogenic trends at regional scales than when a quantity such as surface temperature is averaged over the entire globe (Figure 9.5).

This framework epitomizes the **boundary value problems** (rather than initial value problems) that such GCM experiments are. The key constraint on the model simulation is the *boundary* forcing, or in this case the amount of radiative forcing present. The key constraint of such GCM experiments as those described above is *not* the initial state of the climate system – that is what short-term weather forecasts rely on. In fact, running so many GCMs so many times, each with ever-so-slightly different **initial conditions**, is the idea behind the new wave of large **ensembles**. When an ensemble of simulations by the *same* model is run on the *same* supercomputer, where the only difference is minute changes in the weather on the first day of the model simulation, that is enough of a butterfly effect that each simulation (by the same model, on the same supercomputer) will evolve completely differently in terms of *internal* climate variability. This is another extremely useful source of diversity in model simulations, because averaging the results allows us to tease out what all of these seemingly different predictions actually have in common – that is, the true response to anthropogenic forcing.

9.3   Pathways of Climate Change in the Ocean

In this section, we will examine some of the pathways by which anthropogenic climate change manifests in the global ocean. Focusing on a set of perspectives – heat, volume, ice, and circulation – we will consider the underlying mechanisms linking increases in atmospheric CO_2 concentration and concomitant global warming to changes in the ocean and coupled system, understand the observations and some key sources of observational uncertainty, and evaluate future projections of such changes, including their portfolio of uncertainties. Many of the physical changes that play out within the ocean have the potential to interact with other realms of the Earth system, such as the atmosphere through coupled circulations and the biosphere through the marine carbon cycle, and therefore alter the course of climate change more broadly.

9.3.1 Ocean Heat

Anthropogenic CO_2 is building up in the ocean, but there it does not play the role of greenhouse gas. Ocean warming is driven by a net increase of heat flux across the air–sea interface. In regions where the ocean loses heat to the atmosphere on average (like the western boundary currents), it is losing less heat, and in regions where the ocean gains heat on average (like the equatorial oceans), it is gaining more heat. While the details – such as which terms in the ocean mixed layer heat budget are being altered as a consequence of anthropogenic radiative forcing to yield an overall positive net heat flux anomaly – likely differ by region, the most ubiquitous is that the net longwave radiation is changing such that the longwave radiation emitted upward by the ocean surface is less able to balance the downward longwave radiation emitted in all directions by greenhouse gas molecules in the lower troposphere, as the latter is increasing steadily with the contemporary rise in atmospheric CO_2 concentration. Particularly from an observational perspective, it is useful to consider changes in ocean heat in terms of (a) their expression at the surface (i.e., SST); (b) their contribution to the build-up of heat integrated to great depths; and (c) the vertical structure of temperature changes.

Spotty thermometer records extend back to the 1600s, but our global **instrumental record** (i.e., not including paleoclimate proxies) of SST in the world ocean is only reliable enough for climate studies beginning in the mid-nineteenth century. Several global instrumental reconstructions have been produced by government agencies and climate research groups around the world, and most of them draw on the same bank of raw observations, like the International Comprehensive Ocean-Atmosphere Data Set (ICOADS). This is a collection of global marine observations including SST from ships, buoys, and a few other types of *in-situ* platforms. Most such reconstructions also incorporate the stream of satellite observations of SST beginning in the early 1980s. The ways that these many instrumental reconstructions differ is primarily in the statistical methods with which they take sparse, scattered observations from different places at different times, and produce a nearly complete (in space and time) set of global, gridded fields of SST for every month since around 1850. In other words, how each approaches the task of reducing **sampling error** – the observational uncertainty owing to the fact that the entire ocean surface was not observed all of the time. There are also systematic changes in the SST observing methodology that occurred, which must be taken into consideration when producing a long-term observational record suitable for climate studies. For example, the size of ships changed over time, and near the middle of the twentieth century (around World War II), mariners began measuring SST using engine room intake water rather than a good old bucket. These are sources of **bias error** – the observational uncertainty owing to systematic changes in the observing system over time – and there is more than one way to reduce this uncertainty, including careful adjustments of the affected raw observations prior to attempts to merge with the other data to produce a continuous, globally complete record. In addition to analyses based strictly on observations and statistical techniques to fill the gaps, ocean reanalyses have been developed to achieve the same goal for SST and many other variables. Ocean reanalyses are similar to OGCM simulations, but forced in one way or another (a technique called data assimilation) to adhere to the available observations. The end result may appear similar to an instrumental SST reconstruction, but there is more of an assurance of dynamical consistency between observations and circulation.

Regardless of which instrumental SST reconstruction or ocean reanalysis product is used, the SST over most of the world ocean has been warming by 0.25–1 °C (Figure 8.9), but with some interesting exceptions (see Figure 8.9). One was discussed in Chapter 8 – the cooling trend in the North Atlantic, just southeast of Greenland. It is possible that this is a consequence of reduced poleward heat transport by the AMOC-driven freshening of the high-latitude North Atlantic due to Greenland ice melt. Some instrumental reconstructions suggest that the eastern equatorial Pacific also has been cooling (or at least not warming detectibly), which may be a consequence of a coupled (ocean–atmosphere) response to climate change, which will be discussed in Section 9.3.4. A rather striking feature of the surface temperature trend over the global ocean is the greater warming that has been occurring at high latitudes, a phenomenon called **polar amplification** – we will also return to this in Section 9.3.4 in a discussion of feedbacks. It is also observed that the ocean surface is generally warming less rapidly than the surfaces of the major continental landmasses, which was rather obviously predicted, given the disparity in heat capacity between water and soil, but may have interesting consequences for coastal upwelling systems along the eastern boundaries of the subtropical oceans. One recent study highlighted one of the more direct yet alarming impacts of global and regional SST warming: marine heatwaves. Marine heatwaves put valuable marine ecosystems such as coral reefs and fisheries at risk. Using satellite observations and GCMs, the authors concluded that marine heatwaves have already become more frequent, longer-lasting and more intense, and cover wider areas, and that those trends are projected to continue in the future – just how much depends critically on how many degrees of global warming transpire (Frölicher *et al.*, 2018).

Generating reliable instrumental records of subsurface temperature is similarly challenging, and quite a bit more data-limited than SST. A useful quantity for tracking the overall heat in the ocean is ocean heat content (OHC), effectively temperature integrated from the surface to some depth, multiplied by the density and specific heat of seawater. The resulting quantity is joules of energy lying beneath each square meter of the ocean. Measuring temperature beneath the surface is a bit more of a deliberate act as it cannot easily be done while a ship is moving and is likely impossible to do from satellite. Platforms like the World Ocean Circulation Experiment (repeat ship cruises), Argo (see Chapter 3), and ocean reanalyses now facilitate calculating OHC throughout the world ocean. Consequently, our instrumental record of this quantity is shorter than that of SST but again shows an unequivocal increasing trend in global OHC between 0 and 700 m by about 20×10^{22} J since the late 1950s. Just like we have emitted more than enough CO_2 to explain the rise in atmospheric CO_2 concentration, the rise in atmospheric CO_2 concentration is more than enough to explain the observed warming at the surface of the atmosphere, and here is where about half of that heat is hiding from our surface thermometers – the deep ocean. The amount of excess heat absorbed by the ocean is also variable in time, which causes quite a wave of interesting reactions amid cable news pundits and even some scientists when the global average surface temperature takes a temporary "hiatus" from warming for a few years. In the case of the slowdown of surface warming in the 2000s, it was quickly reconciled as a combination of erroneous data processing and excess heat build-up in the ocean.

With the warming of the ocean originating with surface fluxes, it is logical that the warming would be greatest near the surface and decrease as a function of depth.[3] This is indeed what has been observed since the late 1950s and is predicted in future simulations by GCMs. The consequence of this vertical variation in warming trend is that the vertical stratification of the ocean is increasing – that is, the trend projects onto the background gradient. A stronger vertical stratification in general would act to suppress upwelling, and there are a couple of interesting impacts of that change that we might identify. With all other things being equal, this suppressed upwelling would deliver fewer nutrients to the euphotic (sunlit) zone of the upper ocean and hence reduce photosynthesis driving primary productivity. A recent wide-ranging study of projected changes in marine productivity throughout the twenty-first century using GCMs suggests that this mechanism will indeed drive a decrease in primary productivity across the oceans spanning the tropics to midlatitudes, plus the North Atlantic Ocean (Steinacher et al., 2010). This potential impact of increased vertical stratification – reduced marine primary productivity – is of course a function of many other factors. We will return to the

---

[3] The vertical dependence of temperature trends in the ocean is examined in IPCC (2013), particularly chapter 3, figure 3.1.

question of marine productivity changes in Section 9.3.4 when we consider the impact of increased land–sea thermal contrast noted above on regional circulations. On the other hand, resistance to upwelling and mixing of cold water from below could make tropical storms more intense, since their intense winds on the ocean surface would be less efficient at bringing cooler water to the surface that would serve to weaken the storms. While tropical storms are clearly dependent on the heat energy available in the upper ocean, their impact on society will be exacerbated by another impact of climate change to which increased ocean heat also contributes: sea level rise.

9.3.2   Ocean Volume

Sea level rise is the surface expression of an increase in the volume of the ocean, which can occur at a global scale either by the existing mass of ocean water consuming more volume or by additional mass of water added to the ocean. Both of these mechanisms for sea level rise are mediated by warming and both have contributed roughly equally to the total sea level rise to date. The **steric** component of sea level rise is that due to the thermal expansion of seawater; as the temperature of a fluid increases, its molecules vibrate more and keep a greater distance from neighboring molecules, so the fluid's volume increases while its density decreases. The **eustatic** component of sea level rise is due to additional mass of water added to the ocean. This component is occurring at an increasing rate, particularly from melting the Greenland and Antarctic ice sheets.

The mass of a continental-scale ice sheet is difficult to estimate from the ground. Given the obvious urgency that we know and track such dimensions, a pair of satellites called the Gravity Recovery and Climate Experiment (GRACE) were launched in 2002 to do just that. These satellites orbit the Earth from pole to pole while the Earth spins beneath, and what they are actually measuring is the distance between the twin satellites ("ranging") using a very high-precision laser. The basic idea is that when the leading satellite approaches a swath of the Earth that has more mass, it accelerates forward accordingly, thus increasing the distance between the two satellites ever so slightly. When the trailing satellite approaches the same region of Earth, it accelerates, catches up, and restores the original distance. Every time the GRACE satellites pass over Greenland and Antarctica, they are effectively measuring the mass of ice on those continents. The full GRACE record indicates that Greenland and Antarctica are losing a combined 421 billion metric tons of ice per year, all of which is entering the ocean. These are staggering amounts of water; the Greenland mass loss in 2014 alone was enough to fill Olympic-sized swimming pools stretching from Earth to the moon and back 16 times.

Much like that of SST, the instrumental record of sea level rise extends back with some reliability to the mid-1800s. Prior to the advent of the satellite era, observations of sea level were primarily made by tide gauge stations located along coastlines of continents and islands scattered throughout the world ocean. The obvious source of observational uncertainty associated with this earlier part of the record is one of sampling error; although tide gauges stay put, even an impressive global network of them is only sampling a miniscule fraction of the total surface area of the world ocean. The hope is that those point measurements can be at least partially representative of the complete world ocean. Beginning in the early 1990s, several satellite missions led by European and US agencies have been launched and operated, known as **satellite altimeters**. These are the same satellite altimeters that a physical oceanographer might use to study equatorial Kelvin waves or sea surface height gradients associated with western boundary currents, but when averaged globally they provide a spatially complete (albeit much shorter than the longer-term tide gauge record) record of global mean sea level rise. The basic idea is that the absolute altitude of the satellite above the sea surface is quantified by precisely measuring the time taken for a pulse of microwave energy to be emitted by the satellite, reflect off the surface, and return to the satellite. Since the speed of the microwave pulse is well known, the altitude is simply $c\ \Delta t/2$, where c is the speed of the microwave energy and Δt is the measured return time. The latest satellite altimeters (e.g., Jason 3) have a remarkably small **measurement error** – or uncertainty associated with each altitude estimate by the altimeter instrument – of about 3 cm.[4] When blended together, the observational record of global mean sea level indicates a total rise of about 0.23 m since 1880, with a current rate of sea level rise based on the satellite altimeter record of 3.3 mm per year.

Future projections of sea level rise based on GCMs have considerable scientific uncertainty (Figure 9.6), particularly due to big questions surrounding the *eustatic* component of sea level rise (the physics of thermal expansion is less challenging). How ice sheets such as those on Greenland and Antarctica will behave as the high-latitude regions continue to warm (and to what extent polar amplification will continue to dominate the large-scale expression of surface warming trends) leaves us with a wide range of possible outcomes even under a single assumption about future CO_2 emissions (e.g., RCP4.5 or RCP8.5). Regardless of which RCP becomes the reality of the twenty-first century, GCM simulations associated with the Fifth Assessment Report of the IPCC indicate that global mean sea level rise will proceed at a faster rate than that even over the past few centuries. For example, under RCP4.5 forcing, global mean sea level is projected to rise by 0.47 ± 0.15 m by the end of the twenty-first century (2081–2100) relative to the recent past (1986–2005), and the largest source of (scientific) uncertainty in that projection is that owing to the behavior of the Antarctic ice sheet (Figure 9.6). Coincidentally, the societal uncertainty associated with these global sea level rise projections is similar to

---

[4] The accuracy of the Jason 3 satellite altimeter is 3.3 cm. See https://sealevel.jpl.nasa.gov/missions/jason3 for more information.

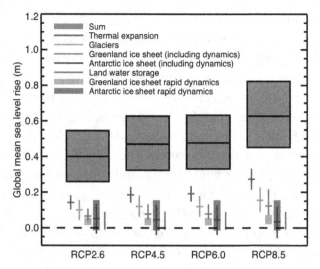

Figure 9.6 Future projections of global mean sea level rise (m) at the end of the twenty-first century (2081–2100) relative to the recent past (1986–2005) under four Representative Concentration Pathways (RCPs) by GCM simulations and process models. Adapted from IPCC AR5 WG1, figure 13.10 (IPCC, 2013). (A black and white version of this figure will appear in some formats. For the color version, please refer to the plate section.)

the aforementioned scientific uncertainty; the difference in the outcome between RCP4.5 and RCP8.5 forcing is also about 0.15 m of sea level rise. Some models indicate that there may be a global mean surface temperature threshold (between 2 °C and 4 °C above pre-industrial) at which a near-complete loss of the Greenland ice sheet may occur over the course of a millennium, raising the global mean sea level by about 7 m. While there remains too little scientific basis to make such risk assessments about catastrophic melting of the Antarctic ice sheet, it is remarkable that Antarctica does hold enough ice to raise global sea level by 60 m!

9.3.3 Sea Ice

Melting sea ice does not contribute to sea level rise, as floating ice has already done so through displacement of liquid water under its own weight. However, this pathway of climate change in the ocean carries important consequences for ocean circulation and global climate, as well as animal life that depends on it for habitat and hunting grounds, such as polar bears. A large field of sea ice such as that in the Arctic Ocean or surrounding Antarctica in the Southern Ocean can be quantified in a few ways, including area or extent, concentration, thickness, and age. Just like temperature and sea level (including contributions from ice sheet melt), the instrumental record of sea ice began with scarce observations[5] (with severe sampling

---

[5] An account of sea ice records prior to the advent of passive microwave satellite measurements in 1979 is given by the National Snow & Ice Data Center (NSIDC); see https://nsidc.org/cryosphere/icelights/2011/01/arctic-sea-ice-satellites.

bias issues) and improved dramatically with the advent of the satellite era. Prior to satellites, sea ice observations were gathered from historical shipping records. Modern satellites primarily use a **passive** microwave remote sensing approach. That is, they only *detect* the microwave emissions from Earth's surface, unlike the afore-mentioned satellite altimeter that beams microwave pulses to Earth and is therefore a form of **active** remote sensing. In the Arctic, the satellite record since 1979 indicates a clear decline in September sea ice extent (when the Arctic sea ice is typically at a seasonal minimum) by about 13 percent per decade. Trends in Arctic sea ice during the seasonal maximum (April) are considerably smaller – only a few percent per year, and sea ice extent trends in the Southern Ocean have also been negligible to date. The strong melting in the Arctic is a potential link in the mystery of why the Arctic region has been warming more rapidly than the rest of the planet (polar amplification); this feedback is discussed in the following section.

Future projections of sea ice melt by GCM simulations present an interesting portfolio of uncertainties in that the difference between forcing scenarios is tanta-mount to a world with Arctic sea ice in the summer from one without (Figure 9.7). Under the lowest forcing scenario (RCP2.6), the spread among models (scientific uncertainty) spans about 0.5–4.5 million km² of sea ice extent at the year 2100.

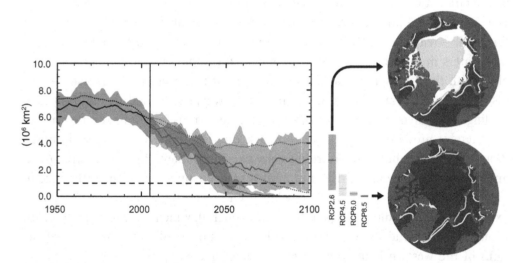

Figure 9.7 Historical simulations and future projections of sea ice extent in the Arctic during September (the seasonal minimum). Shown on the left are the projections under two Representative Concentration Pathways: RCP2.6 and RCP8.5, including the inter-model spread. On the right are the inter-model spreads averaged over 2081–2100 for all four RCPs and Northern Hemisphere maps, indicating the CMIP5 multi-model average over 1986–2005 (white lines) and the CMIP5 multi-model average projections for 2081–2100 (white area). The light blue line and area indicate the results for a subset of CMIP5 models that most closely reproduce the climatological pattern and recent observed trend (1979–2012) of Arctic sea ice extent. Adapted from IPCC AR5 WG1, figure SPM.7 and SPM.8 (IPCC, 2013). (A black and white version of this figure will appear in some formats. For the color version, please refer to the plate section.)

In other words, under that forcing scenario, we are confident that *some* Arctic sea ice will survive the warming in September. Contrast that result with RCP6 or RCP8.5, in which the scientific uncertainty is extremely small and confidently places September Arctic sea ice extent at zero by 2100 (or 2080 in the case of RCP8.5). In considering policy implications for the fate of Arctic sea ice, Alexandra Jahn of the University of Colorado Boulder leveraged an ensemble of GCM future experiments to highlight the fine line between 1.5 °C and 2 °C of global warming in terms of impacts in the Arctic (Jahn, 2018) – the difference was a 30 percent chance of summer ice-free conditions instead of 100 percent certainty of such a change!

9.3.4 Circulations and Feedbacks

While it may be straightforward to anticipate, at least qualitatively, some of the ocean changes like warming, rising, and melting (and acidification) as a direct result of anthropogenic radiative forcing, the impacts on circulations are often more complex, indirect, and therefore uncertain. For a change in the ocean circulation to arise with global warming, it must transpire through the dynamics described in the earlier chapters, especially Chapters 6 and 8 (wind and buoyancy forcing). Moreover, changes in ocean circulation may feed back to the atmospheric circulation and vice versa, yielding *coupled* responses to radiative forcing. The example of a potential ocean circulation that has made a few appearances in this book – reduced AMOC or thermohaline circulation due to freshening of surface waters in the North Atlantic – is one such example where the possible coupling with the atmosphere has been theorized but is not yet well understood. For instance, if the ocean's overturning circulation weakens, would the high-latitude cooling-associated reduced northward ocean heat transport limit the melting of ice, thereby providing a negative feedback on Northern Hemisphere warming? Would the atmosphere's overturning cells like the Hadley circulation strengthen and/or expand poleward (Grise *et al.*, 2019) to make up some of the difference in poleward heat transport? Indeed, there has been substantial evidence of a poleward expansion of the Hadley cells (and tropics in general), which have been linked with changes in the wind-driven ocean circulation as poleward shifts and intensifications of the western boundary currents (Seager and Simpson, 2016; Yang *et al.*, 2016). The exception to the latter is, interestingly, the Gulf Stream, where a weakening AMOC might be pushing that western boundary current the other way.

The amplified surface warming observed in the high-latitude Northern Hemisphere and clear downward trend in Arctic sea ice extent over the past few decades are thought not to be a random coincidence. In fact, this may be one of the most visceral examples of a **positive feedback** loop emerging in the contemporary climate system. The ice–albedo feedback transpires as follows: the surface warms, ice melts, a darker surface is exposed (open water has a much lower albedo than ice), the surface absorbs more solar radiation, warms further, and so on (return to step 1). This mechanism has often been invoked to explain the observed phenomenon of polar amplification, but there are several other feedbacks potentially at play in both the higher latitudes and the tropics, many of which involve clouds and their myriad effects on surface climate. A possible consequence of the polar amplification for changes in the *wind*-driven ocean circulation was identified in a very recent study: An intensification and poleward shift of the western boundary currents *except* for the Gulf Stream where the high-latitude freshening is taking place!

The author has been fortunate to bear witness to – and occasionally participate in – a fascinating debate among tropical climate scientists over the apparent cooling (or at least lack of warming, depending on instrumental SST reconstruction) in the eastern equatorial Pacific Ocean evident over the last century or more. Proposed explanations for this observation can generally fall in one of three different categories: data issues (it's not real; instrumental SST reconstructions are not constrained by enough raw observations in this region), natural variability (this is where ENSO variability plays out, after all), and actual physical mechanisms. One of the physical mechanisms proposed in the mid-1990s by Amy Clement and colleagues at the Lamont–Doherty Earth Observatory of Columbia University directly invokes a coupled circulation response to anthropogenic radiative forcing. The theory is known as the ocean dynamical thermostat (Clement *et al.*, 1996), and goes like this. A more-or-less uniform radiative forcing applied to the whole tropical Pacific Ocean will initially result in stronger SST warming in the western equatorial Pacific Ocean than the east, because upwelling and poleward heat flux divergence is strong in the cold tongue region of the eastern equatorial Pacific. As a result, the zonal SLP gradient along the equatorial atmosphere will strengthen, which will accelerate the easterly trade winds. Stronger easterly trade winds will increase the rates of upwelling, vertical temperature advection, and poleward heat flux divergence via Ekman dynamics, resulting in a *further* cooling of SST in the cold tongue region! Rather than warming, this theory predicts a long-term trend in the mean state of the tropical Pacific sector that bears a strengthened zonal SST gradient and Walker circulation. However, this remains a key regional aspect of global climate change, whose long-term trend observations and GCMs cannot seem to agree on, which is disconcerting considering the outsized role that the tropical Pacific sector plays as an influencer of global climate, not to mention that ENSO variability itself is sensitive to the mean state. Perhaps only time will tell, along with improvements in instrumental records and further developments in GCMs.

Returning to aspects of the global pattern of surface warming that observations and GCMs do seem to agree on – that land will continue to warm faster than the oceans – an interesting hypothesis concerning a coupled circulation change was proposed by Andrew Bakun, a fisheries oceanographer at the US National Oceanic and Atmospheric Administration in Monterey, California. Bakun proposed that strengthened land–sea temperature gradients would lead to similarly strengthened land–sea pressure gradients, which would accelerate the alongshore geostrophic wind, thereby increasing coastal upwelling via offshore Ekman transport along the eastern boundaries of the major subtropical ocean basins (Bakun, 1990; Bakun *et al.*, 2015). The implication is that the increased upwelling would deliver more nutrients to the euphotic zone, increase primary productivity, and potentially have a positive effect at higher trophic levels, like zooplankton and fish. Whether this would be enough to offset reduced nutrient delivery due to increased vertical stratification (as discussed in Section 9.3.1) was and remains unclear. If global warming indeed results in increased upwelling and vertical transport of cool water, a positive feedback is possible, since the SST cooling would further act to enhance the land–sea temperature contrast, and so on.

Finally, we can consider how the ocean may be party to a positive feedback to the anthropogenic radiative forcing itself (not just the warming). Recall that the solubility of CO_2 in seawater decreases with increasing seawater temperature, to the tune of about 3 percent per degree Celsius in colder water (10 °C) and about 2 percent per degree Celsius in warmer water (30 °C). If the upper ocean warms by a few degrees Celsius on a global average, the solubility of CO_2 in the ocean will decrease on the order of 10 percent (Goodwin and Lenton, 2009), meaning the efficiency of the ocean's ability to absorb CO_2 emitted by fossil fuel combustion will decline and more of it will remain in the atmosphere, further warming the atmosphere, further warming the ocean, and so on. This is a troubling carbon cycle–climate feedback on global scale climate, and one recent study estimated its overall impact at about 3 percent of current anthropogenic emissions.

9.4 Closing Remarks

I am always inspired by a short paper written in 1975 by the infamous geophysicist, geochemist, and paleoceanographer Wallace Broecker (Broecker, 1975), whose passing in early 2019 constituted a tremendous loss for the entire scientific community. Professor Broecker, or just "Wally" to his colleagues at the Lamont-Doherty Earth Observatory and around the world, took on the clamoring crowd of pundits eager to dismiss the notion of global warming due to greenhouse gas emission because the temperature had been cooling slightly for a couple of consecutive decades (~1940s to the then-present). Wally made some bold assumptions. First, he assumed aggressive CO_2 emissions over the coming decades – enough to reach 403 ppm by 2010 (which actually happened in 2016). He estimated that half of that would remain in the atmosphere, and that the warming effect of CO_2 was about 0.3 °C per 10 percent increase in concentration. With that "model" in mind, Wally offered a prediction that the cooling trend would reverse within a couple of years, after which Earth would undergo a "pronounced global warming," whereby the global average surface air temperature in ~2013 would be about 1 °C higher than in 1975. Considering the information in front of Wally in 1975, this was a very impressive prediction and was only off by perhaps a few tenths of a degree Celsius, depending on interpretation (Figure 9.8). A very recent study has confirmed that the big GCMs of the variety described in Section 9.2 of this chapter have *also* been quite accurate in their future projections of global temperature change thus far, especially when accounting for the fact that our assumptions of future CO_2 emissions are always imperfect (Hausfather *et al.*, 2019). The predicted climate impacts described above are both serious and credible.

We need more people like that, who are willing to put bold ideas out there and defend them responsibly in the face of noise and doubt. This refers not only to those working in scientific fields like climate or oceanography, but in every sector of the world's affairs, including – especially – those involved in the crucial work of communicating the output of the scientific research enterprise to the general public, stakeholders, and policy makers. This will require the imagination of Wally Broecker, the courage of Greta Thunberg, a broader public education and appreciation of the natural sciences, and a greater diversity of human engagement. Recent studies have demonstrated that the fingerprint of anthropogenic radiative

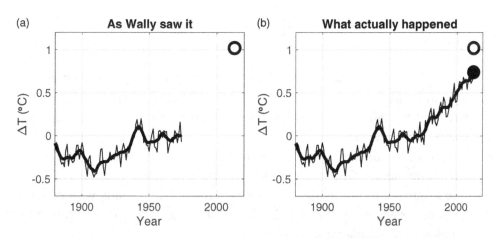

Figure 9.8 Side-by-side portrayals of the global average, annual average surface temperature change beginning in 1880 (ΔT, °C, GISTEMP, thick line indicating the five-year smoothed record). (a) the data available to Wally Broecker in early 1975 (in reality, he probably did not have complete data all the way through 1974 at that time). Wally's prediction of a 1 °C rise in 2013 relative to 1975 is shown by the open circle. (b) The same data but extending from 1975 onward with the actual ΔT in 2013 indicated by the closed circle.

forcing on global temperature and sea level will be detectable for thousands of years (Clark *et al.*, 2016), primarily due to the natural lags built into the ocean, and that with every year that goes by before emissions begin to decline the path to limiting global warming to 1.5 °C or 2 °C becomes less likely or even remotely achievable. The issues outlined in this chapter, and their physical underpinnings detailed in the preceding chapters, truly represents one of the grand challenges of our generation, and it will take more than just scientists to solve it.

For those who *are* considering a career in scientific research in the fields touched on in this book, I assure you that there are more than enough ideas to pursue, mysteries to solve, sources of data to draw upon, and friendly collaborators to work with to keep you busy forever. As I hope you have noticed in your course through this book, one needs little more than a computer and a little creativity to delve quite deeply into the ocean and climate system through a vast array of observations and models covering every corner of the planet. All are welcome in on the fun who enjoy it and are willing to consider that learning never ends. Luckily, I still have a lot to learn.

Further Reading

An excellent book on the scientific basis of modern climate change is *Introduction to Modern Climate Change, Second Edition*, by Andrew Dessler (2016).

Questions

1. Identify two lines of scientific evidence that support the notion that increasing concentrations of greenhouse gases in the atmosphere are causing the observed warming since the nineteenth century. Which line do you think makes a stronger case, and why?
2. What do you think are some key atmospheric variables that an ocean general circulation model (OGCM) requires be provided at its top boundary (i.e., surface forcing) during a simulation? Explain the difference in how those variables are "provided" to the OGCM between standalone OGCM simulations and coupled GCM simulations.
3. Let's assume a weather forecasting agency happens to use as their short-term weather forecast model the same AGCM that constitutes the atmospheric component of a coupled GCM. Despite these similarities across the models used, explain the fundamental differences between a weather forecast and a future climate change simulation.

4.   What are the three general sources of uncertainty in observational climate records? Comment on the relative size or importance of these uncertainties for the climate record of ocean heat content since the late 1950s.

5.   Compare and contrast the blend of uncertainties between GCM future projections of Arctic sea ice and ocean pH. For the latter, you will need to consult the Summary for Policymakers (figure SPM.7) of the Fifth Assessment Report of the IPCC, freely available online.

6.   Several climate system feedbacks were discussed in this chapter. Can you think of one involving clouds? How would it work in terms of cause and effect loop, and is it a positive or negative feedback? Feel free to use your imagination or do a little online research.

7.   Wally Broecker's famous global warming prediction published in a 1975 issue of *Science* was close, but not perfect. How would his simple model predict global temperature to change between "now" (using the 2018 CO_2 concentration of 408.5 ppm) and 2100 under RCP8.5 forcing (930 ppm)? How does that prediction compare to the ensemble of GCMs carrying out that experiment? For the latter, you will need to consult the Summary for Policymakers (figure SPM.7) of the Fifth Assessment Report of the IPCC, freely available online.

8.   Refer to the *Dive into the Data* box featured in this chapter. Read the sample journal article that uses the associated data set, and describe the general role of that data set in the study. Provide a little context, both technical and scientific. What did the authors actually "do" with the data set? What specific scientific insight was enabled by incorporating this data set into the study?

9.   Utilizing *Dive into the Data* Box 2.2, search for evidence supporting or refuting the notion that the so-called Bakun hypothesis has been playing out in eastern boundary upwelling systems over the past 37 years.

10.   Utilizing *Dive into the Data* Boxes 2.2, 3.3, and 9.1, calculate the mean state SST and precipitation biases in the GFDL CM3 and NCAR CCSM4 models relative to observations. Describe the spatial patterns and general magnitudes of any biases you identify. Do you notice any systematic similarities between SST and precipitation biases, model by model? Is it straightforward to say one model is "better" overall than the other in terms of these two variables?

11.   Utilizing *Dive into the Data* Box 9.1, calculate the projected changes in annual mean precipitation in two different regions of the world at global warming levels of 1.5 °C, 2 °C, and 3 °C. How do the two GCMs included in the data set differ in terms of when they reach those global warming levels?

DIVE INTO THE DATA BOX 9.1

Name   Coupled Model Intercomparison Project, Phase 5 (CMIP5)

Synopsis   Global climate models are computer models of the atmosphere, ocean, and other realms such as ice and biosphere. They combine those individual component models to simulate the complete climate system, including the interactions between the aforementioned realms. Approximately 50 different GCMs have been developed and are run on supercomputers at various governmental and academic research institutions around the world. The Coupled Model Intercomparison Project (CMIP), the sixth phase of which is currently afoot, is an effort to coordinate across these institutions to ensure that experiments are conducted in a uniform way and that model output data sets are created and disseminated in a way that is most efficient for the scientists wishing to analyze the results. The entire database is freely available to the public, but a minimum amount of technical understanding of the data and methods for analyzing large volumes of data is usually necessary.

Science   The volume of scientific output facilitated by CMIP5 is enormous and may be categorized as follows: analyses that feed directly into the standard climate change analyses of Assessment Reports produced by the IPCC, and individual climate studies. Analyses feeding directly into IPCC reports include historical attribution experiments, future projections of common regional and global quantities like global mean surface temperature, ocean heat content, sea ice, etc. A very large number of individual climate studies are published each year, focusing on a more specific aspect of climate variability and change, and in cases where a wide range of models is helpful for purposes of quantifying inter-model spread in projections, internal variability in detection and attribution studies, etc., then the CMIP5 base of GCM output data is commonly used.

Figures   9.4, 9.6, 9.7

Version   Phase 5

Variable   Multiple (e.g., surface temperature, precipitation)

Platform   Numerical models

Spatial   Global (ocean, atmosphere, etc.), variable resolution*

Temporal   1860–2100,** monthly averages***

DIVE INTO THE DATA BOX 9.1   (cont.)

Source   https://esgf-node.llnl.gov/search/cmip5
Format   NetCDF (.nc)
Resource   cmip.m; cmip5a.mat; cmip5b.mat; cmip5c.mat; cmip5d.mat
                 (sample data and code provided on publisher website here: www
                 .cambridge.org/karnauskas)

Journal Reference for Data

Taylor, K. E., Stouffer, R., and Meehl, G. An overview of CMIP5 and the
   experiment design. *Bull. Amer. Meteor. Soc.* **93**, 485–498 (2012). DOI:
   10.1175/BAMS-D-11-00094.1.

Sample Journal Article Using Data

Frölicher, T. L., Fischer, E., and Gruber, N. Marine heatwaves under global
   warming. *Nature* **560**, 360–364 (2018). DOI: 10.1038/s41586-018-0383-9.

* A single resolution cannot be defined even for one GCM because different GCMs have
different component models (AGCM, OGCM, etc.) and each of them may have different
resolutions within a GCM. Moreover, an AGCM or OGCM may have spatial resolution that varies
geographically (e.g., higher ocean resolution near the equator is common).
** Historical experiments typically span 1850 or 1860 to 2005, while future experiments begin in
2006. Some future simulations extend to 2300. Other simulations like paleoclimate and control
simulations have different time periods.
*** Daily averages also available.

References

Adler, R. F., Huffman, G., Chang, A., *et al.* The Version-2 Global Precipitation Climatology Project (GPCP) monthly precipitation analysis (1979–present). *J. Hydrometeor.* **4,** 1147–1167 (2003).

Atlas, R., Hoffamn, R., Ardizzone, J., *et al.* A cross-calibrated, multiplatform ocean surface wind velocity product for meteorological and oceanographic applications. *Bull. Amer. Meteor. Soc.* **92,** 157–174 (2011).

Bacastow, R. B., Keeling, C. D., and Whorf, T. P. Seasonal amplitude increase in atmospheric CO_2 concentration at Mauna Loa, Hawaii, 1959–1982. *J. Geophys. Res.* **90,** 10,529–10,540 (1985).

Back, L. E. and Bretherton, C. S. On the relationship between SST gradients, boundary layer winds, and convergence over the tropical oceans. *J. Climate* **22,** 4182–4196 (2009).

Bakun, A. Coastal ocean upwelling. *Science* **247,** 198–201 (1990).

Bakun, A., Black, B., Bograd, S., *et al.* Anticipated effects of climate change on coastal upwelling ecosystems. *Curr. Clim. Change Rep.* **1,** 85–93 (2015).

Barnes, E. A. and Screen, J. A. The impact of Arctic warming on the midlatitude jet-stream: Can it? Has it? Will it? Impact of Arctic warming on the midlatitude jet-stream. *WIREs Clim. Change* **6,** 277–286 (2015).

Bereiter, B., Eggleston, S., Schmitt, J., *et al.* Revision of the EPICA Dome C CO_2 record from 800 to 600 kyr before present. *Geophys. Res. Lett.* **42,** 542–549 (2015).

Bjerknes, J. A possible response of the atmospheric Hadley circulation to equatorial anomalies of ocean temperature. *Tellus* **18,** 820–829 (1966).

Bjerknes, J. Atmospheric teleconnections from the equatorial Pacific. *Monthly Weather Review* **97,** 163–172 (1969).

Bonjean, F. and Lagerloef, G. Diagnostic model and analysis of the surface currents in the tropical Pacific Ocean. *J. Phys. Oceanogr.* **32,** 2938–2954 (2002).

Broecker, W. S. Climatic change: are we on the brink of a pronounced global warming? *Science* **189,** 460–463 (1975).

Brown, J. N., Godfrey, J. S., and Fiedler, R. A zonal momentum balance on density layers for the central and eastern equatorial Pacific. *J. Phys. Oceanogr.* **37,** 1939–1955 (2007).

Burls, N. J., Fedorov, A. V., Sigman, D. M., Jaccard, S. L., Tiedemann, R., and Haug, G. H. Active Pacific meridional overturning circulation (PMOC) during the warm Pliocene. *Science Advances* **3,** 12 pp. (2017).

Caesar, L., Rahmstorf, S., Robinson, A., Feulner, G., and Saba, V. Observed fingerprint of a weakening Atlantic Ocean overturning circulation. *Nature* **556,** 191–196 (2018).

Carranza, M. M. and Gille, S. T. Southern Ocean wind-driven entrainment enhances satellite chlorophyll-a through the summer. *J. Geophys. Res. Oceans* **120**, 304–323 (2015).

Chatterjee, A., Gierach, M., Sutton, A., *et al.* Influence of El Niño on atmospheric CO_2 over the tropical Pacific Ocean: findings from NASA's OCO-2 mission. *Science* **358**, eaam5776 (2017).

Chelton, D. B., Schlax, M., Freilich, M., and Milliff, R. Satellite measurements reveal persistent small-scale features in ocean winds. *Science* **303**, 978–983 (2004).

Cheng, H., Lawrence Edwards, R., Sinha, A., *et al.* The Asian monsoon over the past 640,000 years and ice age terminations. *Nature* **534**, 640–646 (2016).

Chiang, J. C. H. and Vimont, D. J. Analogous Pacific and Atlantic meridional modes of tropical atmosphere–ocean variability. *J. Climate* **17**, 4143–4158 (2004).

Clark, P. U., Shakun, J., Marcott, S., *et al.* Consequences of twenty-first-century policy for multi-millennial climate and sea-level change. *Nature Clim. Change* **6**, 360–369 (2016).

Clarke, A. J. *An Introduction to the Dynamics of El Niño & the Southern Oscillation* (Academic Press, 2008).

Clement, A., Seager, R., Cane, M., and Zebiak, S. An ocean dynamical thermostat. *J. Climate* **9**, 2190–2196 (1996).

Criscitiello, A. S., Das, S., Karnauskas, K., *et al.* Tropical Pacific influence on the source and transport of marine aerosols to west Antarctica. *J. Climate* **27**, 1343–1363 (2014).

de Boyer Montégut, C., Madec, G., Fischer, A. S., Lazar, A., and Iudicone, D. Mixed layer depth over the global ocean: an examination of profile data and a profile-based climatology. *J. Geophys. Res.* **109**, C12003 (2004).

de Jong, M. F. and de Steur, L. Strong winter cooling over the Irminger Sea in winter 2014–2015, exceptional deep convection, and the emergence of anomalously low SST. *Geophys. Res. Lett.* **43**, 7106–7113 (2016).

Del Castillo, C. E., Signorini, S. R., Karaköylü, E. M., and Rivero-Calle, S. Is the Southern Ocean getting greener? *Geophys. Res. Lett.* GL083163 (2019).

Denman, K. L. and Brasseur, G. Couplings between changes in the climate system and biogeochemistry. In *Climate Change 2007: The Physical Science Basis* (ed. Solomon, S., Qin, D., Manning, M., *et al.*) 499–587 (Cambridge University Press, 2007).

Dessler, A. E. *Introduction to Modern Climate Change Second Edition* (Cambridge University Press, 2016).

Dessler, A. E. and Forster, P. M. An estimate of equilibrium climate sensitivity from interannual variability. *J. Geophys. Res. Atmos.* **123**, 8634–8645 (2018).

Di Lorenzo, E., Schneider, N., Cobb, K., *et al.* North Pacific Gyre Oscillation links ocean climate and ecosystem change. *Geophys. Res. Lett.* **35**, L08607 (2008).

Dong, S. and Kelly, K. Heat budget in the Gulf Stream region: the importance of heat storage and advection. *J. Phys. Oceanogr.* **34**, 1214–1231 (2004).

Dong, S., Garzoli, S. L., and Baringer, M. An assessment of the seasonal mixed layer salinity budget in the Southern Ocean. *J. Geophys. Res.* **114**, C12001 (2009).

Dong, S., Gille, S. T., and Sprintall, J. An assessment of the Southern Ocean mixed layer heat budget. *J. Climate* **20**, 4425–4442 (2007).

Drenkard, E. J. and Karnauskas, K. B. Strengthening of the Pacific equatorial undercurrent in the SODA reanalysis: mechanisms, ocean dynamics, and implications. *J. Climate* **27**, 2405–2416 (2014).

Durack, P. J. and Wijffels, S. E. Fifty-year trends in global ocean salinities and their relationship to broad-scale warming. *J. Climate* **23**, 4342–4362 (2010).

Ekman, V. W. On the influence of the Earth's rotation on ocean-currents. *Arkiv for Matematik, Astronomi och Fysik* **2** (11): 1–52 (1905).

EPICA Community Members. One-to-one coupling of glacial climate variability in Greenland and Antarctica. *Nature* **444**, 195–198 (2006).

Feldman, D. R., Collins, W. D., Gero, P. J., Torn, M. S., Mlawer, E. J., and Shippert, T. R. Observational determination of surface radiative forcing by CO_2 from 2000 to 2010. *Nature* **519**, 339–343 (2015).

Florenchie, P., Lutjeharms, J. R. E., Reason, C. J. C., Masson, S., and Rouault, M. The source of Benguela Niños in the South Atlantic Ocean. *Geophys. Res. Lett.* **30** (2003).

Foltz, G. R. and McPhaden, M. J. Seasonal mixed layer salinity balance of the tropical North Atlantic Ocean. *J. Geophys. Res.* **113**, C02013 (2008).

Foltz, G. R., Grodsky, S. A., Carton, J. A., and McPhaden, M. J. Seasonal salt budget of the northwestern tropical Atlantic Ocean along 38°W. *J. Geophys. Res.* **109**, C002111 (2004).

Ford, H. L., Burls, N. J., and Hodell, D. A. Pacific Meridional Overturning Circulation during the Mid-Pliocene warm period (3.264-3.025 Ma). American Geophysical Union Fall Meeting (2019).

Francey, R. J., Allison, C., Etheridge, D., *et al.* A 1000-year high precision record of delta^{13}C in atmospheric CO_2. *Tellus B* **51**, 170–193 (1999).

Frankignoul, C. and Hasselmann, K. Stochastic climate models: Part II. Application to sea-surface temperature anomalies and thermocline variability. *Tellus* **29**, 289–305 (1977).

Frölicher, T. L., Fischer, E. M., and Gruber, N. Marine heatwaves under global warming. *Nature* **560**, 360–364 (2018).

Gierach, M. M., Vazquez-Cuervo, J., Lee, T., and Tsontos, V. M. Aquarius and SMOS detect effects of an extreme Mississippi River flooding event in the Gulf of Mexico. *Geophys. Res. Lett.* **40**, 5188–5193 (2013).

Giese, B. S. and Ray, S. El Niño variability in simple ocean data assimilation (SODA), 1871-2008. *J. Geophys. Res.* **116**, C02024 (2011).

Gill, A. E. Some simple solutions for heat-induced tropical circulation. *Q. J. Royal Met. Soc.* **106**, 447–462 (1980).

Gille, S. T., Stevens, D. P., Tokmakian, R. T., and Heywood, K. J. Antarctic Circumpolar Current response to zonally averaged winds. *J. Geophys. Res.* **106**, 2743–2759 (2001).

Goodwin, P. and Lenton, T. M. Quantifying the feedback between ocean heating and CO_2 solubility as an equivalent carbon emission: CO_2. *Geophys. Res. Lett.* **36** (2009).

Gordon, A. L. *The Role of Thermohaline Circulation in Global Climate Change* (Academic Commons, 1991).

Grise, K. M., Davis, S., Simpson, I., *et al.* Recent tropical expansion: natural variability or forced response? *J. Climate* **32**, 1551–1571 (2019).

Hansen, J., Ruedy, R., Sato, M., and Lo, K. Global surface temperature change. *Rev. Geophys.* **48**, RG4004 (2010).

Harries, J. E., Brindley, H. E., Sagoo, P. J., and Bantges, R. J. Increases in greenhouse forcing inferred from the outgoing longwave radiation spectra of the Earth in 1970 and 1997. *Nature* **410**, 355–357 (2001).

Hasselmann, K. Stochastic climate models: Part I. Theory. *Tellus* **28**, 473–485 (1976).

Hausfather, Z., Drake, H. F., Abbott, T., and Schmidt, G. A. Evaluating the performance of past climate model projections. Geophys. Res. Lett., 10.1029/2019GL085378 (2019).

Hayes, S., McPhaden, M., and Wallace, J. The influence of sea-surface temperature on surface wind in the eastern equatorial Pacific: weekly to monthly variability. *J. Climate* **2**, 1500–1506 (1989).

Hays, J. D., Imbrie, J., and Shackleton, N. J. Variations in the Earth's orbit: pacemaker of the ice ages. *Science* **194**, 1121–1132 (1976).

Honda, M., Yamazaki, K., Tachibana, Y., and Takeuchi, K. Influence of Okhotsk sea-ice extent on atmospheric circulation. *Geophys. Res. Lett.* **23**, 3595–3598 (1996).

Hoskins, B. and Karoly, D. The steady linear response of a spherical atmosphere to thermal and orographic forcing. *J. Atmospheric Sci.* **38**, 1179–1196 (1981).

Hu, C., Lee, Z., and Franz, B. Chlorophyll *a* algorithms for oligotrophic oceans: a novel approach based on three-band reflectance difference. *J. Geophys. Res.* **117**, (2012).

Huang, B., Thorne, P., Banzon, V., *et al.* Extended Reconstructed Sea Surface Temperature, Version 5 (ERSSTv5): upgrades, validations, and intercomparisons. *J. Climate* **30**, 8179–8205 (2017).

IPCC. *Climate Change 2013: The Physical Science Basis. Working Group I Contribution to the Fifth Assessment Report of the Intergovernmental Panel on Climate Change* (Cambridge University Press, 2013).

Jacox, M. G., Moore, A. M., Edwards, C. A., and Fiechter, J. Spatially resolved upwelling in the California Current System and its connections to climate variability. *Geophys. Res. Lett.* **41**, 3189–3196 (2014).

Jahn, A. Reduced probability of ice-free summers for 1.5 °C compared to 2 °C warming. *Nature Clim. Change* **8**, 409–413 (2018).

Jayne, S. R. and Bogue N. M. Air-deployable profiling floats. *Oceanography* **30**, 29–31 (2017).

Kanamitsu, M., Ebisuzaki, W., Woollen, J., *et al.* NCEP-DOE AMIP-II Reanalysis (R-2). *Bull. Amer. Meteor. Soc.* **83**, 1631–1644 (2002).

Karnauskas, K. B. and Li, L. Predicting Atlantic seasonal hurricane activity using outgoing longwave radiation over Africa: African OLR and Atlantic hurricanes. *Geophys. Res. Lett.* **43**, 7152–7159 (2016).

Kay, J. E., Deser, C., Phillips, A., *et al.* The Community Earth System Model (CESM) Large Ensemble Project: a community resource for studying climate change in the presence of internal climate variability. *Bull. Amer. Meteor. Soc.* **96**, 1333–1349 (2015).

Keeling, C. D., Bacastow, R. B., Bainbridge, A. E., *et al.* Atmospheric carbon dioxide variations at Mauna Loa Observatory, Hawaii. *Tellus* **28**, 538–551 (1976).

Kilpatrick, K. A., Podestá, G., Walsh, S., *et al.* A decade of sea surface temperature from MODIS. *Remote Sens. Environ.* **165**, 27–41 (2015).

Knight, J. R., Allan, R., Folland, C., Veillina, M., and Mann, M. A signature of persistent natural thermohaline circulation cycles in observed climate. *Geophys. Res. Lett.* **32**, L20708 (2005).

Koblinsky, C. J., Hildebrand, P., LeVine, D., *et al.* Sea surface salinity from space: science goals and measurement approach. *Radio Sci.* **38**, 8064–8067 (2003).

Kouketsu, S., Kaneko, I., Kawano, T., *et al.* Changes of North Pacific Intermediate Water properties in the subtropical gyre. *Geophys. Res. Lett.* **34**, L02605 (2007).

L'Heureux, M. L., Takahashi, K., Watkins, A. B., *et al.* Observing and predicting the 2015/16 El Niño. *Bull. Amer. Meteor. Soc.* **98**, 1363–1382 (2017).

Lau, K. M. and Kim, K. M. Cooling of the Atlantic by Saharan dust. *Geophys. Res. Lett.* **34** (2007).

Le Quéré, C., Andrew, R., Friedlingstein, P., *et al.* Global carbon budget 2018. *Earth Syst. Sci. Data* **10**, 2141–2194 (2018).

Li, J., Scinocca, J., Lazarre, M., *et al.* Ocean surface albedo and its impact on radiation balance in climate models. *J. Climate* **19**, 6314–6333 (2006).

Liebmann, B. and Smith, C. Description of a complete (interpolated) outgoing longwave radiation dataset. *Bull. Amer. Meteor. Soc.* **77**, 1275–1277 (1996).

Lindzen, R. and Nigam, S. On the role of sea-surface temperature-gradients in forcing low-level winds and convergence in the tropics. *J. Atmospheric Sci.* **44**, 2418–2436 (1987).

Lisiecki, L. E. and Raymo, M. E. A Pliocene–Pleistocene stack of 57 globally distributed benthic $\delta^{18}O$ records. *Paleoceanography* **20** (2005).

Liu, J., Bowman, K., Schimel, D., *et al.* Contrasting carbon cycle responses of the tropical continents to the 2015–2016 El Niño. *Science* **358**, eaam5690 (2017).

Loeb, N. G., Doelling, D., Wang, H., *et al.* Clouds and the Earth's Radiant Energy System (CERES) Energy Balanced and Filled (EBAF) Top-of-Atmosphere (TOA) Edition-4.0 data product. *J. Climate* **31**, 895–918 (2018).

Lovelock, J. E. Hands up for the Gaia hypothesis. *Nature* **344**, 100–102 (1990).

Lüthi, D., Le Floch, M., Bereiter, B., *et al.* High-resolution carbon dioxide concentration record 650,000–800,000 years before present. *Nature* **453**, 379–382 (2008).

Maes, C., Picaut, J., and Belamari, S. Importance of the salinity barrier layer for the buildup of El Niño. *J. Climate* **18**, 104–118 (2005).

McCarthy, G. D., Smeed, D. A., Johns, W. E., *et al.* Measuring the Atlantic Meridional Overturning Circulation at 26°N. *Prog. Oceanogr.* **130**, 91–111 (2015).

McPhaden, M. and Taft, B. Dynamics of seasonal and intraseasonal variability in the eastern equatorial Pacific. *J. Phys. Oceanogr.* **18**, 1713–1732 (1988).

McPhaden, M. J., Busalacchi, A., Cheney, R., *et al.* The Tropical Ocean–Global Atmosphere (TOGA) observing system: a decade of progress. *J. Geophys. Res.* **103**, 14,169–14,240 (1998).

Mitas, C. M. and Clement, A. Has the Hadley cell been strengthening in recent decades? *Geophys. Res. Lett.* **32**, L03809 (2005).

Moum, J. N., Perlin, A., Nash, J. D., and McPhaden, M. J. Seasonal sea surface cooling in the equatorial Pacific cold tongue controlled by ocean mixing. *Nature* **500**, 64–67 (2013).

Munk, W. H. and Carrier, G. F. The wind-driven circulation in ocean basins of various shapes. *Tellus* **2**, 160–167 (1950).

Nagura, M. and McPhaden, M. J. The dynamics of zonal current variations in the central equatorial Indian Ocean. *Geophys. Res. Lett.* **35**, L23603 (2008).

Nagura, M. and McPhaden, M. J. Zonal momentum budget along the equator in the Indian Ocean from a high-resolution ocean general circulation model. *J. Geophys. Res. Oceans* **119**, 4444–4461 (2014).

Newman, M., Alexander, M., Ault, T., *et al.* The Pacific Decadal Oscillation, revisited. *J. Climate* **29**, 4399–4427 (2016).

Newman, S. M., Smith, J. A., Glew, M. D., Rogers, S. M., and Taylor, J. P. Temperature and salinity dependence of sea surface emissivity in the thermal infrared. *Q. J. R. Meteorol. Soc.* **131**, 2539–2557 (2005).

North Greenland Ice Core Project members. High-resolution record of Northern Hemisphere climate extending into the last interglacial period. *Nature* **431**, 147–151 (2004).

O'Neill, L. W., Chelton, D. B., and Esbensen, S. K. Observations of SST-induced perturbations of the wind stress field over the Southern Ocean on seasonal timescales. *J. Climate* **16**, 2340–2354 (2003).

Past Interglacials Working Group of PAGES. Interglacials of the last 800,000 years. *Rev. Geophys.* **54**, 162–219 (2016).

Paulson, C. and Simpson, J. Irradiance measurements in upper ocean. *J. Phys. Oceanogr.* **7**, 952–956 (1977).

Qiao, L. and Weisberg, R. The zonal momentum balance of the equatorial undercurrent in the central Pacific. *J. Phys. Oceanogr.* **27**, 1094–1119 (1997).

Qu, T., Gao, S., and Fukumori, I. What governs the North Atlantic salinity maximum in a global GCM? *Geophys. Res. Lett.* **38** (2011).

Rahmstorf, S. Ocean circulation and climate during the past 120,000 years. *Nature* **419**, 207–214 (2002).

Rahmstorf, S. Thermohaline circulation: the current climate. *Nature* **421**, 699–699 (2003).

Ren, L., Speer, K., and Chassignet, E. P. The mixed layer salinity budget and sea ice in the Southern Ocean. *J. Geophys. Res.* 116, C08031 (2011).

Reynolds, R., Rayner, N., Smith, T., Stokes, D., and Wang, W. An improved in situ and satellite SST analysis for climate. *J. Climate* 15, 1609–1625 (2002).

Roemmich, D. and Gilson, J. The 2004–2008 mean and annual cycle of temperature, salinity, and steric height in the global ocean from the Argo Program. *Progress in Oceanography* 82, 81–100 (2009).

Rossby, C.-G. Planetary flow patterns in the atmosphere. *Q. J. R. Meteorol. Soc.* 66, 68–87 (1940).

Rossow, W. B. and Schiffer, R. A. ISCCP cloud data products. *Bull. Amer. Meteor. Soc.* 72, 2–20 (1991).

Rubino, M., Etheridge, D., Trudinder, C., *et al.* A revised 1000 year atmospheric δ^{13}C-CO$_2$ record from Law Dome and South Pole, Antarctica. *J. Geophys. Res. Atmos.* 118, 8482–8499 (2013).

Sagan, C. *Pale Blue Dot: A Vision of the Human Future in Space* (Random House, 2004).

Sandström, J. W. Dynamicsche Versuche mit Meerwasser. *Annalen der Hydrographie under Martimen Meteorologie*, 36, 6–23 (1908).

Sarachik, E. S. and Cane, M. A. *The El Niño-Southern Oscillation Phenomenon* (Cambridge University Press, 2010).

Sato, K., Suga, T., and Hanawa, K. Barrier layers in the subtropical gyres of the world's oceans. *Geophys. Res. Lett.* 33, L08603 (2006).

Schmitt, R., Bogden, P., and Dorman, C. Evaporation minus precipitation and density fluxes for the north-Atlantic. *J. Phys. Oceanogr.* 19, 1208–1221 (1989).

Schmitz, W. J. *On the World Ocean Circulation. Volume I: Some Global Features/North Atlantic Circulation* (Woods Hole Oceanographic Institution, 1996).

Seager, R. and Simpson, I. R. Western boundary currents and climate change: currents and climate change. *J. Geophys. Res. Oceans* 121, 7212–7214 (2016).

Small, R. J., Bryan, F. O., Bishop, S. P., and Tomas, R. A. Air–sea turbulent heat fluxes in climate models and observational analyses: what drives their variability? *J. Climate* 32, 2397–2421 (2019).

Smeed, D. A., Josey, S. A., Beaulieu, C., *et al.* The North Atlantic Ocean is in a state of reduced overturning. *Geophys. Res. Lett.* 45, 1527–1533 (2018).

Sprintall, J. and Tomczak, M. Evidence of the barrier layer in the surface layer of the tropics. *J. Geophys. Res.* 97, 7305–7316 (1992).

Steinacher, M., Joos, F., Frölicher, T., *et al.* Projected 21st century decrease in marine productivity: a multi-model analysis. *Biogeosciences* 7, 979–1005 (2010).

Stephenson, D. B., Wanner, H., Brönnimann, S., and Luterbacher, J. *The History of Scientific Research on the North Atlantic Oscillation* (American Geophysical Union, 2003).

Stevens, W. K. Scientist at work: Wallace S. Broecker; iconoclastic guru of the climate debate, *New York Times*, March 17, 1998, section F, page 1 (1998).

Stommel, H. The westward intensification of wind-driven ocean currents. *Trans. AGU* **29**, 202 (1948).

Stommel, H. The abyssal circulation. *Deep Sea Res.* **5**, 80–82 (1958).

Stommel, H. and Arons, A. B. On the abyssal circulation of the world ocean: II. An idealized model of the circulation pattern and amplitude in oceanic basins. *Deep Sea Res. (1953)* **6**, 217–233 (1959a).

Stommel, H. and Arons, A. B. On the abyssal circulation of the world ocean: I. Stationary planetary flow patterns on a sphere. *Deep Sea Res. (1953)* **6**, 140–154 (1959b).

Sud, Y. C. and Walker, G. K. Simulation errors associated with the neglect of oceanic salinity in an atmospheric GCM. *Earth Interactions* **1**, 1–19 (1997).

Sverdrup, H. U. Wind-driven currents in a baroclinic ocean: with application to the equatorial currents of the Eastern Pacific. *PNAS* **33**, 318–326 (1947).

Sverdrup, H. U., Johnston, M. W., and Fleming, R. H. *The Oceans: Their Physics, Chemistry, and General Biology* (Prentice-Hall, Inc., 1942).

Taylor, K. E., Stouffer, R. J., and Meehl, G. A. An overview of CMIP5 and the experiment design. *Bull. Amer. Meteor. Soc.* **93**, 485–498 (2012).

Trenberth, K. and Caron, J. Estimates of meridional atmosphere and ocean heat transports. *J. Climate* **14**, 3433–3443 (2001).

Trenberth, K. E., Branstator, G., Karoly, D., *et al.* Progress during TOGA in understanding and modeling global teleconnections associated with tropical sea surface temperatures. *J. Geophys. Res.* **103**, 14,291–14,324 (1998).

Trenberth, K. E., Smith, L., Qian, T., Dai, A., and Fasullo, J. Estimates of the global water budget and its annual cycle using observational and model data. *J. Hydrometeor.* **8**, 758–769 (2007).

Trenberth, K. E., Fasullo, J. T., and Kiehl, J. Earth's global energy budget. *Bull. Amer. Meteor. Soc.* **90**, 311–324 (2009).

van Vuuren, D. P., Edmonds, J., Kainuma, M., *et al.* The representative concentration pathways: an overview. *Clim. Change* **109**, 5–31 (2011).

Vásquez-Bedoya, L. F., Cohen, A. L., Oppo, D. W., and Blanchon, P. Corals record persistent multidecadal SST variability in the Atlantic Warm Pool since 1775 AD. *Paleoceanography* **27** (2012).

Vellinga, M. and Wu, P. Low-latitude freshwater influence on centennial variability of the Atlantic thermohaline circulation. *J. Climate* **17**, 4498–4511 (2004).

Vihma, T. Effects of Arctic sea ice decline on weather and climate: a review. *Surv. Geophys.* **35**, 1175–1214 (2014).

Wacongne, S. Dynamical regimes of a fully nonlinear stratified model of the Atlantic equatorial undercurrent. *J. Geophys. Res.* **94**, 4801 (1989).

Walin, G. On the relation between sea-surface heat flow and thermal circulation in the ocean. *Tellus* **34**, 187–195 (1982).

Walker, G. T. Correlation in seasonal variability of weather: VIII. A preliminary study of world weather. *Mem. India Meteorol. Dept.* **24**, 75–131 (1923).

Wallace, J., Mitchell, T., and Deser, C. The influence of sea-surface temperature on surface wind in the eastern equatorial Pacific: seasonal and interannual variability. *J. Climate* **2**, 1492–1499 (1989).

Wang, C., Xie, S. P., and Carton, J. A. *Earth's Climate: The Ocean-Atmosphere Interaction* (American Geophysical Union, 2004).

Wang, J. and Bras, R. L. A new method for estimation of sensible heat flux from air temperature. *Water Resour. Res.* **34**, 2281–2288 (1998).

Xie, S.-P. Satellite observations of cool ocean–atmosphere interaction. *Bull. Amer. Meteor. Soc.* **85**, 195–208 (2004).

Xie, S.-P. and Philander, S. G. H. A coupled ocean–atmosphere model of relevance to the ITCZ in the eastern Pacific. *Tellus A* **46**, 340–350 (1994).

Yan, Y., Li, L., and Wang, C. The effects of oceanic barrier layer on the upper ocean response to tropical cyclones. *J. Geophys. Res. Oceans* **122**, 4829–4844 (2017).

Yang, H., Lohmann, G., Wei, W., *et al.* Intensification and poleward shift of subtropical western boundary currents in a warming climate. *J. Geophys. Res. Oceans* **121**, 4928–4945 (2016).

Yu, L., Jin, X., and Weller, R. A. Multidecade global flux datasets from the Objectively Analyzed Air-Sea Fluxes (OAFlux) Project: latent and sensible heat fluxes, ocean evaporation, and related surface meteorological variables. Woods Hole Oceanographic Institution, OAFlux Project Technical Report. OA-2008-01 (Woods Hole Oceanographic Institution, 2008).

Yu, X. and McPhaden, M. Dynamical analysis of seasonal and interannual variability in the equatorial Pacific. *J. Phys. Oceanogr.* **29**, 2350–2369 (1999).

Zhang, L., Karnauskas, K. B., Weiss, J. B., and Polvani, L. M. Observational evidence of the downstream impact on tropical rainfall from stratospheric Kelvin waves. *Clim. Dyn.* **50**, 3775–3782 (2018).

Zhao, M., Held, I. M., Lin, S.-J., and Vecchi, G. A. Simulations of global hurricane climatology, interannual variability, and response to global warming using a 50-km resolution GCM. *J. Climate* **22**, 6653–6678 (2009).

Index

Printed in the United States
by Baker & Taylor Publisher Services